# Vibration measurement and analysis

# Vibration Measurement and Analysis

**J. D. Smith** MA PhD CEng MIMechE
University Lecturer, Engineering Department, University of Cambridge

**Butterworths**
London Boston Singapore Sydney Toronto Wellington

PART OF REED INTERNATIONAL P.L.C.



First published 1989



**British Library Cataloguing in Publication Data**
Smith, J. D.
Vibration measurement and analysis.
1. Mechanical vibration. Engineering aspects
I. Title
620.3

ISBN 0–408–04101–3

**Library of Congress Cataloging in Publication Data**
Smith, James D. (James Derek), 1934–
Vibration measurement and analysis.

Bibliography: p. 162
Includes index.
1. Vibration. I. Title.
TA355.S53 1989 620.3 88–37586
ISBN 0-408-04101-3

Typeset in Great Britain by EJS Chemical Composition, Bath
Printed and bound in Great Britain by Butler & Tanner, Frome, Somerset

# Contents

# Preface

This book is intended for the engineer who is starting in the field of vibration measurement and analysis techniques and who wishes to gain a broad understanding of the many rival approaches and their advantages and, of course, their disadvantages.

I have deliberately avoided the use of mathematics as far as is possible since the complexities of the maths rarely assist understanding of the physical processes and it is not necessary for the user of equipment to know the exact details of a particular method, merely its validity, range and limitations.

Equipment has not usually been referred to by trade names since, although the principles are general, different companies hold sway in different territories. Despite this I hope that a secure base in fundamentals will allow a practising engineer to make suitable choices.

Analysis is currently dominated by the frequency analysis approach which, although very valuable in the majority of cases, is deceptive and destructive when irregular intermittent impacts are encountered. I hope that the reader will gain some insight into the choice between frequency and time domain methods.

This book owes much to friends in academic life and in industry who have taught me and expanded my experience, and to my wife Rona for her patience and support during a long gestation.

Chapter 1

# Introduction

## 1.1 Reasons for measurement

It comes as a surprise to many engineering graduates that it is still necessary to measure vibrations, due to the mistaken belief that it is possible to calculate 'everything' on a computer. The reality is very different and it is worth while considering the reasons for measuring.

One reason lies in the fallibility of predictions of stiffnesses. At first sight this is surprising, since finite element techniques with fine resolution can give the stiffness of an element with great accuracy, but problems usually occur when elements of a machine meet. It is even difficult to predict the performance of a joint with high-precision flat surfaces, whose bolts have been tightened under laboratory conditions, since the stiffness is dependent on very local distortions. Under normal production conditions any joint, slide-way or bearing gives unpredictable and, equally important, sometimes unrepeatable results. Large systems also present problems since vibrations within the structure and significant stress wave propagation times require massive computational effort; it is thus frequently cheaper to test than to predict.

Damping is, however, the main uncertainty in most engineering problems, since damping alone controls the heights of the resonance peaks. This is extremely important. The mechanics of damping in machines is still not understood and claims that damping in normal structures can be predicted should be treated with scepticism. Typically, it is possible to predict the frequency of a resonance to within 10%, but even given some experience of similar machines, estimates of damping can easily be out by a factor of 4. This is perhaps not too surprising when tests on a particular machine can vary by 50% from one day to the next.

Another major reason for making measurements on equipment occurs when the amount of excitation is not known. The most common case occurs in the balancing of rotors where the cause of vibration is an initial imbalance that remains variable despite attempts at balancing before assembly. Automobile wheels are a particular (rather crude) example where a system that is balanced initially deteriorates due to wear and must be rebalanced at intervals. Hydraulic systems almost inevitably generate pressure pulses as each cylinder in a pump or motor encounters a change of pressure.

On a production line the combination of variation in excitation levels and variation in system response can lead to successive products that are apparently identical but that give up to 15 dB difference in noise levels. Only testing on a vibration rig will detect the small percentage of extremely noisy components, in order that money is not wasted by, e.g., building a gearbox into a vehicle and then having to remove it due to excess noise.

Modern equipment can require very low levels of vibration for a variety of reasons; at one end of the scale a gearbox transmitting 10 000 kW may require a vibration level of less than one-tenth of a micron to avoid sonar detection, while an electron microscope can give a distorted picture if relative vibrations of one-thousandth of a micron (10 Å) are present.

## 1.2 Single degree of freedom systems

Many of the measuring instruments used for vibration work are themselves an approximation to a single degree of freedom vibrating system, thus it is worth while keeping the characteristics of such systems (and the inherent assumptions made) in mind when considering their limitations.

The classic idealisation of a single degree of freedom system is shown in Figure 1.1; it is slightly confusing that a single degree of freedom system is also called a second order system, referring to the second differential in the equation. The inherent assumptions are that the mass is perfectly rigid, the spring is linear and massless, and the damping is linear and dependent only upon velocity. The analysis infers that the spring and damping forces have no effect upon the base, which is rigid. The acceleration of the mass, neglecting steady effects due to gravity, is given by the sum of the forces

$$m\ddot{\mathbf{x}} = -s\mathbf{x} - \lambda\dot{\mathbf{x}} + \mathbf{F},$$

i.e.

$$m\ddot{\mathbf{x}} + \lambda\dot{\mathbf{x}} + s\mathbf{x} = \mathbf{F},$$

or

$$\frac{\ddot{\mathbf{x}}}{s/m} + \frac{\ddot{\mathbf{x}}}{s/\lambda} + \mathbf{x} = \frac{\mathbf{F}}{s}.$$

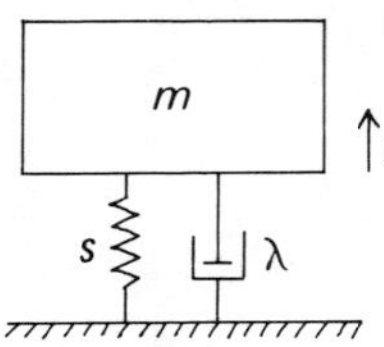


**Figure 1.1** Single degree of freedom idealisation with applied vibrating force **F**; $\lambda$ is the viscous damper coefficient and $s$ is the spring stiffness and $m$ is the mass

The solution of this differential equation will have a complementary function that corresponds to a starting transient in the system and a particular integral that corresponds to the excitation from the right-hand side of the equation.

In vibration, unlike some other branches of engineering, there are usually only two excitations of interest, the impulse and steady sinusoidal vibration; these are linked mathematically and all other vibration encountered in machines can be derived from either of these two particular cases. Most work in vibration is expressed in terms of the frequency response, i.e. the response to a steady sinusoid after initial transients have decayed, and this approach will be followed in this book.

Taking first the simplest case when the base is fixed and a steady oscillating force **F** $\cos \omega t$ is applied, the equation becomes

$$\frac{\ddot{\mathbf{x}}}{\omega_n^2} + 2c\frac{\dot{\mathbf{x}}}{\omega_n} + \mathbf{x} = \frac{\mathbf{F}}{s}\cos \omega t, \text{ where } \begin{array}{l} \omega_n^2 = s/m \\ 2c = \omega_n \lambda / s \end{array} \qquad \text{(a)}$$

The equation may be solved by any of the standard methods for differential equations, omitting the transient term, or by making the engineering deduction that if the forcing is at frequency $\omega$ radians per second, the response must be at the same frequency but a phase angle $\phi$ lagging behind. The in-phase ($\cos \omega t$) and quadrature ($\sin \omega t$) terms must each balance in the equation.

The final result is

$$\mathbf{x} = \frac{(\mathbf{F}/s)\cos(\omega t - \phi)}{\{[1-(\omega/\omega_n)^2]^2 + [2c(\omega/\omega_n)]^2\}^{1/2}}\,; \tan\phi = \frac{2c(\omega/\omega_n)}{1-(\omega/\omega_n)^2}$$

The natural frequency $\omega_n$ is sometimes called the undamped frequency of the system or the 90° phase frequency, while the non-dimensional factor $c$ is the fraction of the critical damping which would just make the system non-oscillatory; this corresponds to equal real roots of the differential equation.

As it is usually only the amplitude that is of interest we can regard the solution as the quasi-static or very low frequency

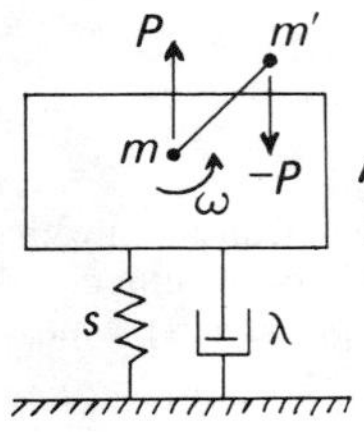


**Figure 1.2** Excitation by an out-of-balance mass $m'$ at a radius $e$ on a single degree of freedom (vertical) system

response $\mathbf{F}/s$ multiplied by a dynamic magnification factor which is $([1-(\omega/\omega_n)^2]^2+[2c\omega/\omega_n]^2)^{-1/2}$.

The case of an unbalanced rotor with main mass $m$ and imbalance of $m'$ at radius $e$, as in Figure 1.2, gives the equations

$$m\ddot{\mathbf{x}} = -s\mathbf{x} - \lambda\dot{\mathbf{x}} + P$$

$$m'(\ddot{\mathbf{x}} - e\omega^2 \cos \omega t) = -P$$

(where $P$ is the force between $m$ and $m'$) for vertical motion only. Combining and dividing by $s$ gives

$$\frac{\ddot{\mathbf{x}}}{\omega_n^2} + 2c\frac{\dot{\mathbf{x}}}{\omega_n} + \mathbf{x} = \frac{m'}{s}e\omega^2 \cos \omega t = \frac{m'e}{(m+m')} \cdot \left(\frac{\omega}{\omega_n}\right)^2 \cos \omega t \qquad \text{(b)}$$

where now

$$\omega_n^2 = s/(m+m')$$

The solution to this equation follows for the case above and is

$$|\mathbf{x}| = \frac{m'e}{m+m'}\left(\frac{\omega}{\omega_n}\right)^2 \frac{1}{\{[1-(\omega/\omega_n)^2+[2c(\omega/\omega_n)]^2\}^{1/2}}.$$

Again, this is the force $m'e\omega^2$ divided by the spring stiffness $s$ to give a quasi-static deflection, multiplied by the same dynamic magnification factor as before. The phase angle expression is the same as for case (a).

Neither of the above cases correspond to the 'seismic excitation' condition which is relevant for vibration transducers, where the idealisation in Figure 1.3 shows the excitation as vertical vibration of the ground. The equation of motion is

$$m\ddot{\mathbf{x}} = -s(\mathbf{x}-\mathbf{y}) - \lambda(\dot{\mathbf{x}} - \dot{\mathbf{y}}),$$

so

$$\frac{\ddot{\mathbf{x}}}{\omega_n^2} + 2c\frac{\dot{\mathbf{x}}}{\omega_n} + \mathbf{x} = \frac{\lambda}{s}\dot{\mathbf{y}} + \mathbf{y}. \qquad \text{(c)}$$

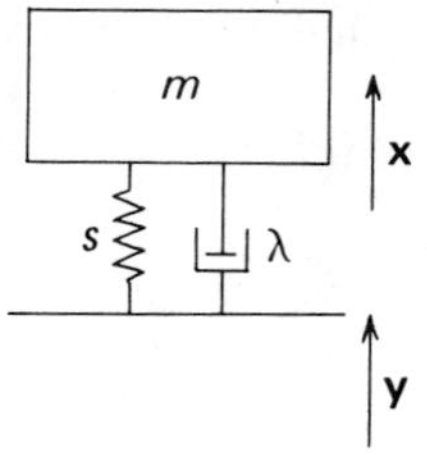


**Figure 1.3** Excitation due to ground vibrating vertically with amplitude **y**. The resulting vibration **x** is an absolute vibration, *not* relative to **y**

Here the 'exciting force' is $\lambda\dot{\mathbf{y}} + s\mathbf{y}$ and the solution for **x** is

$$|\mathbf{x}| = |\mathbf{y}|\left[1+2c\left(\frac{\omega}{\omega_n}\right)^2\right]^{1/2}$$

$$\times\left\{\left[1-\left(\frac{\omega}{\omega_n}\right)^2\right]^2+\left[2c\frac{\omega}{\omega_n}\right]^2\right\}^{-1/2}$$

and $\phi$, the phase angle, is given by

$$\tan\phi = 2c\left(\frac{\omega}{\omega_n}\right)^3 \Big/ \left[1-(1-4c^2)\left(\frac{\omega}{\omega_n}\right)^2\right].$$

However, we are not interested in the displacement of the mass since a transducer such as a piezoelectric accelerometer works by measuring the force through the crystal, i.e. the supporting force which is equal to $m\ddot{\mathbf{x}}$.

$$|m\omega^2\mathbf{x}| = |m\omega^2\mathbf{y}|\left[1+2c\left(\frac{\omega}{\omega_n}\right)^2\right]^{1/2}$$

$$\times\left\{\left[1-\left(\frac{\omega}{\omega_n}\right)^2\right]^2+\left(2c\frac{\omega}{\omega_n}\right)^2\right\}^{-1/2}.$$

The dynamic magnification factor is similar to that for cases (a) and (b) but has an extra term in the numerator due to the damping force.

A few transducers, such as seismic velocity pickups, can be idealised as shown in Figure 1.3, but working above their natural frequency so that the mass $m$ remains stationary and the system then measures the relative movement or velocity between 'ground' and mass, i.e. $(\mathbf{x}-\mathbf{y})$ or its differential. The analysis can be carried out by deducing **x** from equation (c), as above, and then carrying out a vector subtraction of **x** from **y**, bearing in mind that **x** lags **y** by the angle $\phi$. However, the arithmetic of this approach is longwinded.

It is simpler to change the variable in the original equation by putting $\mathbf{z} = \mathbf{x} - \mathbf{y}$ then

$$m\ddot{\mathbf{x}} = -s(\mathbf{x} - \mathbf{y}) - \lambda(\dot{\mathbf{x}} - \dot{\mathbf{y}})$$

$$m(\ddot{\mathbf{z}} + \ddot{\mathbf{y}}) = -s\mathbf{z} - \lambda\dot{\mathbf{z}}$$

or

$$m\ddot{\mathbf{z}} + \lambda\dot{\mathbf{z}} + s\mathbf{z} = -m\ddot{\mathbf{y}} = m\omega^2 \mathbf{y} \cos \omega t.$$

This is of the same form as case (b) and so has the solution

$$|\mathbf{z}| = \mathbf{y}\left(\frac{\omega}{\omega_n}\right)^2 \left\{\left[1 - \left(\frac{\omega}{\omega_n}\right)^2\right]^2 + \left[2c\frac{\omega}{\omega_n}\right]^2\right\}^{-1/2}.$$

As $\omega$ becomes large this approximates to $\mathbf{y}$, as expected, since $\mathbf{x}$ becomes small.

The above analyses apply equally well to torsional systems with the sole change that the original equation is of the form

$J\ddot{\theta}$ = sum of all torques due to springs, damping and excitation,

where

$J$ is the rotational moment of inertia of the system.

## 1.3 Two degrees of freedom

A system with two degrees of freedom, idealised as in Figure 1.4, gives two resonances as might be expected, but also shows the phenomenon of an antiresonance. As antiresonances often occur in testing and are very important it is worth while examining their behaviour.

Damping complicates the arithmetic, thus omitting damping:

$$m\ddot{\mathbf{x}} = -s_1\mathbf{x} - s_2(\mathbf{x} - \mathbf{y}) + \mathbf{F} \cos \omega t,$$

$$M\ddot{\mathbf{y}} = -s_3\mathbf{y} + s_2(\mathbf{x} - \mathbf{y}).$$

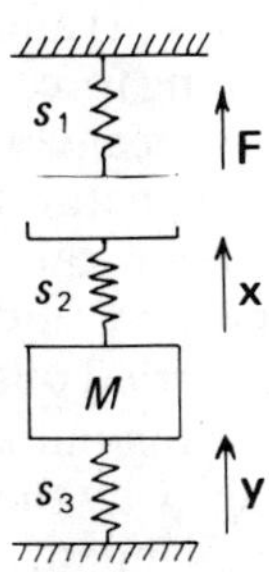


**Figure 1.4** Idealisation of two degrees of freedom with no damping

We now indulge in a typical engineering manoeuvre of the simultaneous assumptions that we have no damping so that we can write down the equations above and also that we have sufficient damping to eliminate starting transients. Any response of the system must then be at frequency $\omega$ and, since there is no damping, must be exactly inphase or antiphase with the excitation. This allows us to substitute $\ddot{\mathbf{x}} = -\omega^2 \mathbf{x}$ and $\ddot{\mathbf{y}} = -\omega^2 \mathbf{y}$ since $A \cos \omega t$ differentiates twice to $-\omega^2 A \cos \omega t$. This gives

$$(-m\omega^2 + s_1 + s_2)\,\mathbf{x} = s_2^2 \mathbf{x}/(-M\omega^2 + s_2 + s_3) + \mathbf{F} \cos \omega t$$

and so

$$x = \frac{(s_2 + s_3 - M\omega^2)\,\mathbf{F} \cos \omega t}{(s_1 + s_2 - m\omega^2)(s_2 + s_3 - M\omega^2) - s_2^2}.$$

Resonances occur at the values of $\omega^2$, which make the denominator zero, and an antiresonance occurs when $\omega^2 = (s_2 + s_3)/M$.

The physical explanation of the antiresonance is that the mass $M$ is moving with exactly the correct amplitude (and phase) such that the force in spring $s_2$ is equal and opposite to the applied force, i.e. $s_2 \mathbf{y} + \mathbf{F} = 0$.

A general observation with many degrees of freedom is that resonances and antiresonances alternate as indicated in Figure 1.5, and the presence of a resonance of a small part of a machine can often be detected only by its effect on reduction of amplitude at the forcing position.

Vibration absorbers use extra masses carefully tuned to important frequencies as one method of reducing resonant effects in structures but it is usual then to add damping; Den Hartog [1] describes the applications and the theory of optimising dampers.

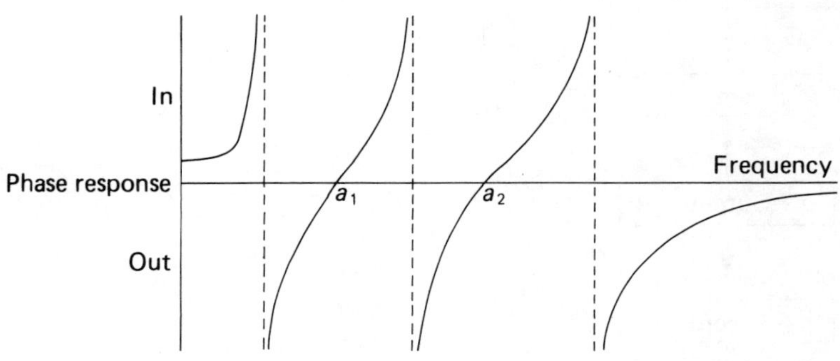


**Figure 1.5** Typical response of system with many degrees of freedom, showing alternating resonances and antiresonances (at $a_1$ and $a_2$)

Chapter 2

# General measurement requirements

## 2.1 The measurement chain

Carrying out a vibration investigation will involve several stages whose sequence is illustrated in Figure 2.1.

As displacement, velocity, acceleration, or force cannot be measured directly, the first stage is for a transducer to convert the mechanical change into change of an electrical parameter such as resistance, voltage, charge, inductance or capacitance. It is of course possible to have non-electrical systems but these are so rarely used now that we do not consider them here.

The second stage, usually electronic in some form or another, converts the parameter change into an electrical voltage (or current) which, as in the case of a strain gauge bridge or photodiode, may only be of the order of microvolts.

An intermediate third stage may then be required to buffer, amplify, demodulate, integrate, add or subtract signals to give a voltage that is ideally proportional to the original displacement,

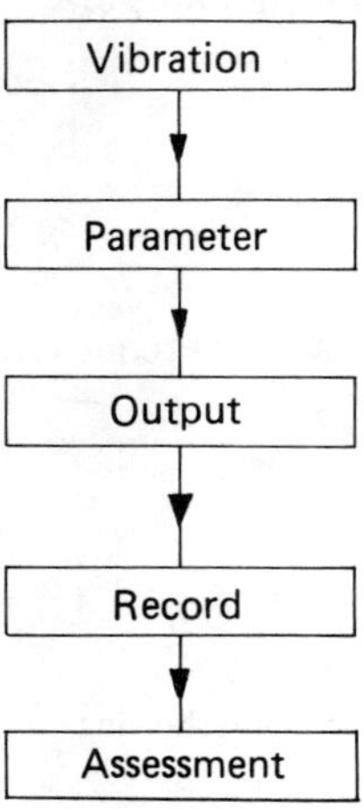


**Figure 2.1** Sequence of stages in measuring vibration

etc., of sufficient amplitude that it is well above noise levels and ready for viewing or recording.

The next stage is normally to record the signal so that the basic data is available for reference, and finally the signal may be analysed by methods to be discussed later.

At each stage of the chain it is important that the information is not distorted or destroyed, but the most critical stage is the initial transducer stage, and thus most attention focusses on the possible errors that may be introduced and which criteria govern the choice of transducer.

## 2.2 Transducer criteria

Choosing one transducer and its associated electronics for a particular application involves a process of elimination of unsuitable transducers; this is in contrast to, say, choosing a car. The process of selection is governed by the problems and inadequacies of transducers, which renders manufacturers' literature somewhat unreliable because, inevitably, salesmen emphasise the positives and gloss over the problems. It is thus necessary to read sales literature with a very critical eye and to assess the requirements in relation to the limits set by linearity, frequency range, mechanical impedance matching and, occasionally, electrical impedance matching.

## 2.3 Linearity and discrimination

There are some types of transducer, such as accelerometers, which are inherently linear so that output is directly proportional to input, but distance transducers are usually inherently non-linear and much effort must be put into linearising the required signal.

It is theoretically possible, now that cheap computing power is available, to use non-linear devices and to correct the results from an accurate calibration curve. This method is only used if there is no alternative because it is a relatively slow process and complicates testing in the field. The normal procedure is to demand linearity from the manufacturer and buy elsewhere if this is not provided.

The definition of linearity is rather more complicated than it might appear and a bland '0.2% linearity' may be deceptive. Linearity is usually defined in relation to full range on the device, so that a transducer capable of measuring over 25 mm (1 inch) may be correct to within 50 $\mu$m (0.002″) and produce unacceptable errors if

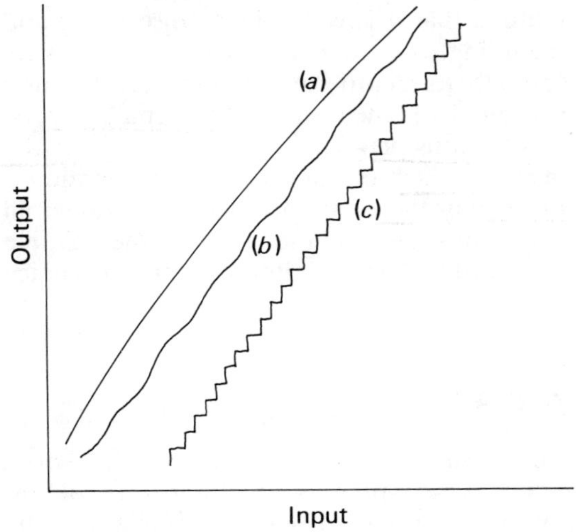


**Figure 2.2** Non-linearity types: (*a*) is curved; (*b*) shows a short wavelength ripple; and (*c*) is a stepped or staircase extreme version of (*b*)

there is, say, an important fifth harmonic in the motion with a total amplitude of only 0.25 mm (0.010″).

Figure 2.2 indicates the types of non-linearity that may occur; (*a*) is usually relatively unimportant but (*b*) and (*c*) have serious effects for small vibrations. These problems can occur either as a result of coarse digitisation at the recording stage, or they may be due to an inherent characteristic such as a single turn of wire on a badly designed wire-wound linear potentiometer or inductive transducer.

A full specification of non-linearity will involve not only the full range specification but some form of point-to-point local figure. A typical case of a circular grating might well read: any line to any other line 5 seconds of arc; line-to-line within 5° movement, accuracy 2 seconds of arc; line to neighbouring line, 0.5 seconds of arc. If linearity is critical, the information should be given as a plot or table of variation from a straight line, since the actual input to output plot cannot show small variations which may be less than 0.1%.

The other terms that may be encountered are discrimination and consistency or repeatability. Discrimination refers to a device that has an output of the form illustrated in Figure 2.2(*c*) and gives the step height; most modern pickups have infinite discrimination so that any movement, however small, will give a change in output.

However, with contacting devices there is often a combination of friction and elasticity which gives a small dead-zone analagous to the 'sticktion' in a dial gauge, and this gives a finite level of discrimination of the order of 1 $\mu$m (40 $\mu$inches). Consistency refers to the ability to repeat an output when the system returns to exactly the same mechanical position; lack of consistency may be due to mechanical effects such as 'sticktion' or hysteresis, or may be due to drift effects in the electronics.

## 2.4 Frequency range

Figure 2.3 is a sketch of a typical accelerometer and electronics response curve. The curve has a central plateau where the output is proportional to acceleration, regardless of the frequency of vibration. Below this range the sensitivity decreases due to electronic limitations and above this range the accelerometer resonates mechanically and gives a response that varies as shown.

The extent of the range may either be quoted by specifying the frequencies that correspond to ±10% variation from the response within the 'flat' section, or by specifying the 3 dB point at the bottom end and the resonance frequency. The latter method is usually adopted but is slightly deceptive since the 30% reduction in amplitude at the 'half-power point' is not usually permissible and it is necessary to restrict the top frequency to about 40% of the resonant frequency if the sensitivity is to remain within 20% for a low damped system. The best method of specifying the range is to give the calibration curve.

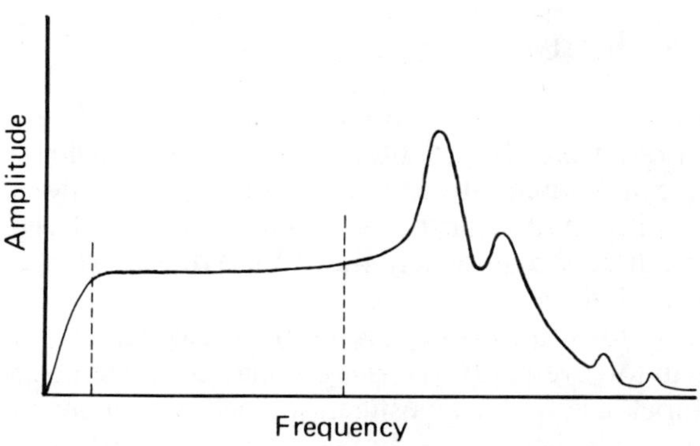


**Figure 2.3** Typical accelerometer response curve

As with linearity, it is possible to compensate by computer for variations in frequency response, and this approach can extend the usable range by an octave at each end, but it is usually easier to use a different transducer than to compensate the results.

A problem can arise at the ends of the range if it is important that the phase of a signal is recorded correctly since phase errors can occur even when amplitudes are correct. On the assumption that at the low frequency end the system behaves as a first order system, then at a frequency of three times the 3 dB frequency the reduction in amplitude sensitivity is cos (arctan 1/3) which at 5% is not important, but phase distortion of arctan 1/3 (i.e. 18°) can be important for control systems or waveform work.

More troublesome is the effect that occurs with systems that work with relatively low frequency AC carriers. Typical of this type of system are some strain gauge bridges and some displacement transducers that use an AC carrier to give improved stability, consistency and freedom from noise. The final stage of an AC system involves some form of rectification and smoothing which introduces time delays. A system working with an 8 kHz carrier would normally have a bandwidth of up to 2 kHz and an inherent time delay of 0.5 ms. This does not sound much but corresponds to 180° phase shift at 1000 Hz; even at 100 Hz the 18° phase shift must be allowed for when phase is important. Systems whose carrier frequency is above 50 kHz will not usually give significant phase shift, but some linear variable displacement transducers working on carriers well below 1000 Hz can give large phase shifts at 10 Hz due to excessive smoothing.

## 2.5 Cross-axis sensitivity

Cross-axis sensitivity is a measure of the error signal produced by a transducer if it is vibrated exactly at right-angles to its working axis. This criterion is relevant when vibration occurs in two directions simultaneously, as with shaft whirling or where there is a large vibration in the feed direction but where the motion of importance is gap variation in the slideway.

For many vibrations the motion of interest is the major vibration, and cross-axis sensitivity (even of 10%) does not matter, but if mode shapes of a rigid block are under investigation it is convenient to locate the instantaneous centre by finding directions of zero vibration and here cross-axis errors are very important.

## 2.6 Electrical impedance matching

Impedance matching in the electrical sense is relatively straightforward and involves ensuring that the output characteristics at one stage are suitable for the input characteristics of the next stage. Devices such as photocells used with gratings must be followed by a low input impedance device, otherwise system frequency response is severely limited due to slow decay of the charges generated. Similarly, accelerometers with long leads prefer a low impedance charge amplifier to a conventional voltage amplifier.

Later in the instrumentation chain it is usual for outputs to be voltages from low impedance sources and for inputs to have high impedance, typically 1 MΩ, so that outputs do not 'load' or affect the input.

An exception occurs in the case of logic or counting circuits that are based on transistor-transistor-logic or TTL. Such systems cause problems because they are so very different in concept from analogue systems that silly results are frequent and interfacing presents problems. The most commonly used system works between 0- and 5-volt rails and has an input which, if left, will normally sit high at a voltage of between 3 and 5 V. Operation of the circuit occurs when a current of about 1.6 mA is drawn from the input to pull its voltage below 2 V and switch the input 'low'. The output of a device in the 'low' state is capable of 'pulling down' a current of the order of 16 mA and so controlling several inputs, but an output in the 'high' condition can supply very little current, about $40 \times 10^{-9}$ A being typical.

If the output is fed to an analogue device such as a tape recorder with an input impedance of 100 kΩ, then at 5 V the recorder would require $5 \times 10^{-5}$ A. But only one-thousandth of this is available from a TTL 'output'. The problem can be overcome easily with an extra current supply from the 5 V rail, as indicated in Figure 2.4.

Transition from analogue to TTL circuits is usually straightforward since many analogue circuits are based on operational

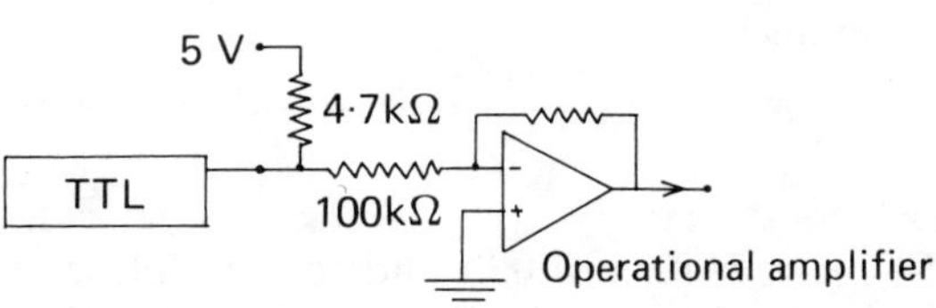


**Figure 2.4** Transition from TTL to an analogue circuit. The 'pull-up' resister supplies the extra current needed for the amplifier input

amplifiers similar to the 741 type and are capable of draining 1.6 mA easily. It is often advisable to put a 4.7 V Zener diode across the TTL input to protect it from excess voltage. A problem can arise if an analogue vibration signal, swinging symmetrically about 0 V, is connected both to a recorder and to a frequency counter. Some frequency counters, designed with TTL signals in mind, are protected by Zener diodes, and when the input signal is negative the diode operates and allows large currents to flow from earth to the input signal. The currents exceed the output capability of the (741) amplifier which saturates and gives a distorted signal; this is irrelevant where counting only is required but it is serious if the signal is also being recorded. A resistance of the order of 5 kΩ in series with the counter will solve the problem.

## 2.7 Mechanical matching

Problems of mechanics can influence the choice of transducer due to contact resonances, excessive mass or added damping. Excessive mass is fairly obvious since it is easy to see whether, say, the 30 g mass of an accelerometer is significant relative to the structure being tested. Trouble usually occurs on thin panels, such as automobile bodies, where a thickness of less than 1 mm is typical; the 'effective mass' seen at the excitation point in a large plate varies with frequency when vibrated transversely and can be calculated [1].

An infinite plate of thickness $h$, modulus $E$, density $\varrho$ and Poisson's ratio $\nu$ has a response at the driving point given by

$$\frac{F}{v} = 8\left[\frac{E}{(1-\nu^2)} \cdot \frac{h^3}{12} \cdot \varrho h\right]^{1/2}$$

which is

$$8h^2[E\varrho/12(1-\nu^2)]^{1/2}.$$

The effective mass $M_e$ is given by $\mathbf{F} = M_e \cdot (\omega v)$, since acceleration is $|\omega v|$, and thus is

$$M_e = \frac{8h^2}{\omega}[E\varrho/12(1-\nu^2)]^{1/2}$$

For steel $E$ is $210 \times 10^9$, $\varrho$ is 7840, $\nu$ is 0.3, and so for a plate of thickness 1 mm, i.e. $10^{-3}$ m, and a frequency of 160 Hz ($\omega = 1000$), the effective mass is only 98 g. This corresponds to the mass of a disc of the same thickness and radius 63 mm. Use of a standard accelerometer of mass 20 g would not be suitable for frequencies above about 50 Hz. Ideally, transducers should not be more than about 5% of the effective vibrating mass.

Mounting of accelerometers on steel objects is simple with a small magnet but the additional mass may cause trouble or the magnetic pull be insufficient to hold against high acceleration levels.

Rather more difficult to quantify is the damping involved if a transducer is hand-held; pressure is needed to maintain contact so that energy is transmitted to vibrate fingers or hand. As with mass, this is not a problem with heavy objects but may affect testing of thin panels or light plastics.

Contact resonances are by far the most common mechanical problem; they are associated with elasticity of the contact between transducer and test object. This is likely to be of the order of $10^7$ N/m ($5 \times 10^4$ lbf/in), so that a transducer mass of about 1 kg would resonate at 500 Hz, but an accelerometer of mass 25 g would not resonate till 3 kHz. Use of a flat surface for contact gives very much higher stiffness, but of an unknown amount. Minor surface variations will give large differences in stiffness and so it is advisable to smear a small amount of thick grease, petroleum jelly or beeswax between the mating surfaces to increase the contact stiffness; resonant frequencies of the order of 30 kHz can be achieved with small transducers either bolted on or stuck on [2].

## 2.8 Conditioning

The analogue signal from a transducer may be suitable for direct use or may need additional amplification, filtering, addition or subtraction, integration or buffering. Simple operational amplifier circuits will perform these jobs in most cases, although high performance filters with slopes of more than 12 dB/octave are best bought in. Details of circuits and their limitations are discussed in Chapter 6.

## 2.9 Viewing or recording

The final stage of the instrumentation chain is to view and/or record the signal. Simple viewing of the vibration signal on an oscilloscope is much underrated but is very valuable as a check. There are virtually no frequency limitations with an oscilloscope and it is sufficiently linear for viewing purposes.

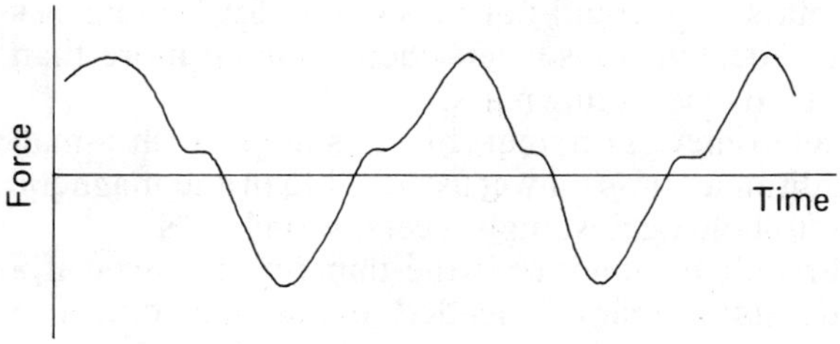


**Figure 2.5** Signal distortion due to a non-linearity such as backlash

Direct viewing of the trace is valuable because it will show very quickly if the signal is exceeding the expected voltage levels or if signal distortion is occurring due to saturation at any point in the chain. Occasionally a force signal will be distorted, as shown in Figure 2.5, indicating some form of play or rattle. Viewing can also be useful if the signal is erratic or the vibration of interest occurs regularly at a particular point of an operating cycle, since the bursts of vibration show where attention should be concentrated for recording or analysis.

Recorders, whether analogue or digital must cover the frequency range of interest so it is important to determine which range is relevant in advance of tests. A pressure wave may have a rise time of only 20 $\mu$s and thus if a digital approach is used, a sample rate of about one megacycle is required (or for tape recording a frequency range of 500 kHz). Audible noise investigations range up to 20 kHz, thus either a 20 kHz tape recorder should be used or about a 50 kHz sample rate digitally.

## 2.10 Signal to noise ratio

At any stage in the measurement chain the signal to noise ratio (S/N) of the voltage signal may be critical. This parameter is most likely to give trouble if there is a wide range of amplitudes present at differing frequencies, and consideration should always be given to avoiding signal to noise ratio problems by, say, using velocity instead of acceleration, or vice versa (see Section 6.5).

The definition of a signal to noise ratio is not an exact process since the 'noise' may be due to different causes and hence will have different characteristics. The main types of background noise are:

1. Thermal noise: due to random vibration of atoms at finite temperatures. Typically, a few $\mu$V peak to peak at amplifier

inputs. As it is random it is broadband and is usually specified as an energy density $V^2/Hz$ or as V per $(Hz)^{1/2}$. A typical figure for an instrumentation amplifier might be $5\,nV/(Hz)^{1/2}$; in a band between, say, 950 Hz and 1050 Hz we would then expect to see 50 nV only, but peak voltages would be at least an order higher.

2. Mains pickup: occurs mainly at harmonics of mains frequency except when unshielded leads are used. This type of noise, though irritating, is not troublesome provided equipment is not running synchronously.
3. Tape noise: only relevant for tape recording, it is the well-known 'hiss' that tape produces due to the finite size of the 'magnets' that record the signal. It occurs randomly but tends to increase with frequency and so is troublesome when directly recording high frequencies. There may also be problems due to interference or variable drive speeds but these are unlikely on high quality equipment. High frequency vibrations associated with the electronics can also occur and should be checked for, either by using a pure calibration sine wave or by using a steady DC input and frequency-analysing the output.
4. Digitisation noise: can take several forms but is associated with the finite size of the smallest bit. Spurious or exaggerated components can be introduced, so that any observed vibration amplitude of less than the least significant bit (0.4% of full scale for 8-bit) should be treated with caution.
5. Demodulation noise: associated with the demodulation of AC systems. Typically with an 8 kHz carrier there may be significant noise at subharmonics; some noise at 1 kHz and 2 kHz is common for such a system.

As the types of noise vary so much and the signals under investigation vary in type, it is very difficult to get a realistic assessment in general terms. Particular types of noise may be detected simply by a zero-input test but other types, such as digitisation noise, are highly dependent on the type of signal present.

Chapter 3

# Types of transducer

## 3.1 Classifications

Transducers may be referred to (apart from trade names) either by their electrical characteristic, e.g. a capacitance pickup, or by what they measure, e.g. a velocity pickup, or the measurement may be specified in more detail, e.g. a non-contact relative velocity pickup. No method of classification is satisfactory but as a resistive device may be an accelerometer, a force transducer or a relative displacement transducer, specifying resistance is not informative and it is customary to refer to 'a displacement transducer' or 'an accelerometer'.

A slight complication occurs because it is normal to use 'an accelerometer' to measure velocity or displacement with the necessary integrations carried out in the electronics. Similarly, 'strain gauges' may be used to measure displacement or force.

A simple but fundamental division is between devices that measure absolute vibration and those that measure relative vibration; the second group can then be subdivided into contact and non-contact systems. The 'absolute' vibration pickups subdivide into the majority, which employ single-degree-of-freedom systems below resonance and use the force required to accelerate the mass as a measure of acceleration; less common are the systems that have a single degree of freedom system working above its (low) resonance frequency. These 'seismic' pickups rely on an internal mass that essentially remains stationary since excitation is at high frequency, and so measurement of relative displacement or velocity between mass and external case gives the absolute movement of the case.

## 3.2 Contact relative transducers for displacement

The simplest transducer of this type has a DC voltage across a linear resistance that is wire-wound helically, or a carbon track similar to a conventional potentiometer, or a conducting plastic strip. A spring loaded wiper makes contact and the output voltage is directly proportional to displacement.

Wire-wound devices tend to suffer from resolution problems since there may be a step in the output at every wire, typically 50 $\mu$m (0.002″), unless great care has been taken to smooth the transfer from wire to wire. In the past this type of transducer tended to have a limited track life, but improvements in materials have now greatly improved performance and it is relatively easy to obtain large strokes. An advantage of the resistive type of linear transducer lies in its freedom from the phase distortions that the AC types suffer and so it is popular for fields such as hydraulics research where large strokes of the order of 0.25 m are needed but where phase matters. These pickups have virtually no frequency limits and have good linearity for long strokes but are less suitable for short strokes. Some friction is involved and masses to be moved are comparable with rival systems.

DC systems as described above are used relatively rarely, particularly for short strokes such as the 2 mm required for an 'electronic dial gauge'. The system generally used is based on the principle sketched in Figure 3.1 and uses an alternating voltage carrier. The coils *A* and *C* are similar and are fed with the same alternating current directed so that the magnetic fields oppose each other in the centre at *B*. If coil *B* is exactly in the centre no voltage will be induced, but if it is moved towards *A* there will be a greater voltage induced by *A* than the opposing voltage induced by *C* and there will be an output voltage in *B* which will be proportional to distance from the symmetrical position.

In practice a separate movable coil *B* would be bulky, heavy, rather flimsy, and there would be some difficulty in restraining it in line with coils *A* and *C*. So the three coils are fixed rigidly on a tube and variation in the magnetic coupling of *A* to *B* and *B* to *C* is achieved by moving a small ferrite slug along the centre of the tube. In the sketch the slug is indicated in the position where there is a maximum coupling of *B* to *A* and thus is at one end of its design stroke. The ferrite slug is typically only about 2 mm in diameter and as the operating rod is of similar diameter, the total mass to be moved is small and sufficiently robust to stand vibration.

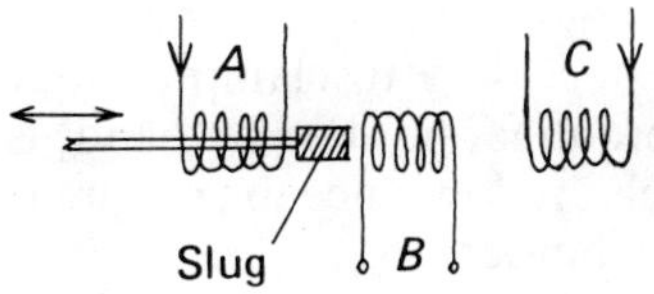


**Figure 3.1** Principle of linear variable distance transducer with ferrite slug providing variable coupling between the centre and outer coils

This type of inductive transducer is sometimes called a linear variable distance transducer (LVDT) and is particularly popular for intermediate distances in the 1–100 mm region. When carefully made it is highly linear and as it is an AC device it gives a stable and repeatable output. Like all AC systems there is phase distortion; at 8 kHz carrier frequency this is typically 18° per 100 Hz.

## 3.3 Relative velocity contact transducers

This type of device uses the property that a wire which cuts lines of force in a magnetic field will generate a voltage proportional to the field intensity and the velocity of the wire.

Physically the transducer consists of a coil that moves along its axis in a constant radial magnetic field provided by a permanent magnet; it is the reverse of the moving coil operation in a standard loudspeaker. Figure 3.2 illustrates a typical design; the probe is spring loaded onto the vibrating surface and the main mass of the magnetic circuit and case is either hand-held or clamped to a vibration-free base.

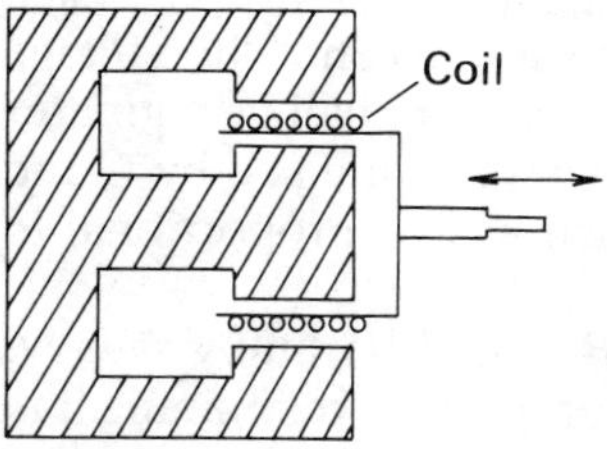


**Figure 3.2** Principle of moving coil velocity transducer using cylindrical coil in a radial magnetic field

Many coil velocity pickups are reasonably robust and have good linearity; the effective mass of the coil and suspension assembly is low, typically only a few grams, and they are easy to use. Frequency range is limited at the top end by contact resources where the coil mass vibrates on the Hertzian elastic deflections; this occurs at about 2 kHz for normal designs but special miniature types used for bearing testing have coils of only 3 mm diameter and achieve contact resonances above 10 kHz. A major limitation is that they measure relative velocity and so it is necessary to clamp a fairly bulky and heavy case rigidly. The alternative, holding by hand, is not usually satisfactory if vibration levels are low since the output is dominated by the muscle tremors of the holder.

## 3.4 Strain gauges

These are a particular form of relative displacement transducer not usually associated with vibration measurement but, rather, with stress determination. Some transducers do, however, use strain to measure movement by winding fine wire between stationary and moving pins in a pressure gauge, as shown in Figure 3.3. Variation of wire length and the corresponding Poisson reduction in diameter gives a resistance change.

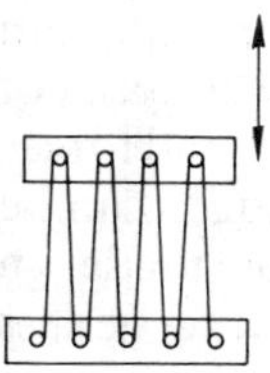

**Figure 3.3** Strain gauge, unbonded, for distance measurement. The fine wire is anchored and insulated by ruby pins

More commonly, the wire is wound onto a paper or plastic backing and the complete gauge is bonded onto a surface to detect stress. Stresses are rarely of interest in vibration work, but it is sometimes convenient to use the bending stresses at the base of a cantilever as a guide to movement at the tip. When applying vibration the most convenient method of measuring force is to use strain gauges on a link between vibrator and test piece.

Strain gauges have a very wide frequency range, to above 50 kHz and, being linear, they meet all the standard criteria. The disadvantages arise from the very low level of the output signal and the major effects of temperature. Typical figures for a gauge are:

resistance 120 Ω ± 0.3;
gauge factor, i.e. ratio of resistance change to strain 2.0;
temperature coefficient $3 \times 10^{-5}$/°C.

The high value of the temperature coefficient means that a temperature change of 30° C gives a resistance change equivalent to one-third of the yield in mild steel. To overcome this an unstressed dummy gauge can be used for balance; if temperature gradients exist in the component the dummy gauge must be very close to the working gauge. The easiest way to achieve this is to use a crossed pair of gauges, as illustrated in Figure 3.4; a pair of gauges set at 45°, as shown, can also be used as a direct measure of shear stress since shear stress in a given direction is equivalent to pure tension and compression at 45°.

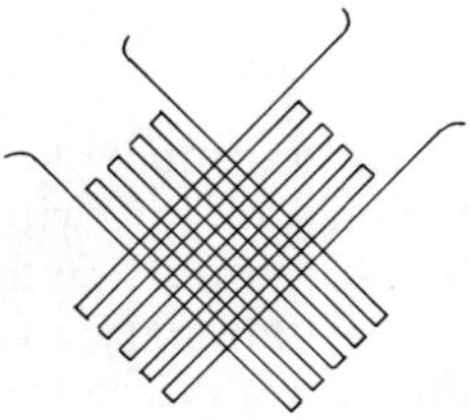

**Figure 3.4** Pair of strain gauges for temperature compensation or shear measurement

All six components of force and torque can be measured in a 'top-hat' dynamometer, as illustrated in Figure 3.5; the rigid top and bottom sections are needed to give even stresses in the thin cylindrical shell that carries the gauges. Up to 24 crossed pairs of gauges set at 45° or axially are connected to give moments and forces as required, with low interaction and high rigidity. Gauges can be made to work at high temperatures, and they can be made on semiconductors to give high sensitivities but these tend to be less stable than metallic types.

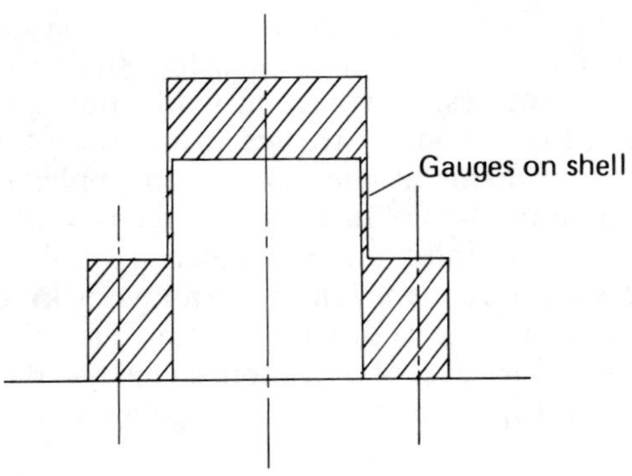


**Figure 3.5** Top-hat dynamometer. The thin (1 mm) shell carries the gauges while very rigid top and bottom sections give even stresses

## 3.5 Non-contact relative displacement transducers

This family of transducers has a large number of variants despite the relatively restricted use of non-contact types; the main area of application lies in the measurement of relative movement between rotating shafts and bearing housings for monitoring or investigation of bearing performance and shaft whirling.

Systems use a fairly high frequency alternating current to a probe that is of the order of 10 mm diameter positioned less than 2 mm away from a metallic surface. Variation of probe distance alters either the capacitance, the inductance or the eddy current losses. Capacitance types require the surface to be earthed and are affected by oil mist changing the dielectric constant, but they are not affected by variations in target conductivity or magnetic properties. Both the inductive and loss types depend on the generation of eddy currents in the target surface; the nearer the target the stronger the induced currents and thus the greater the inductance and resistive losses. Variation of resistivity of the surface alters the eddy currents and, in order to avoid spurious apparent distance variations, it is therefore important to calibrate against the correct material and to check that a rotating shaft surface has consistant properties.

All three types are fundamentally non-linear so the manufacturers have to compensate for their characteristics or use the probes over very restricted ranges. In practice they usually succeed in achieving about 1% linearity, which is more than adequate for normal vibration work.

Frequency characteristics of the transducers are good because they will work down to DC, allowing easy calibration; and since the alternating frequencies used are high the frequency response is good (normally 10 kHz) and can range up to 50 kHz. There is, however, likely to be some phase distortion at high frequencies and there may be slight drift problems when attempting to measure mean gap.

Since there is no contact with the vibrating surface there are no mechanical loading problems. The only problem is to ensure that the 'stationary' probe is fixed rigidly; this typically involves about a 10 mm tapped hole. The probe mass is low, of the order of 30 g.

Sensitivity or resolution is difficult to define and is often not given; for one of the most popular transducers it is quoted as about 1 $\mu$m. This is more than adequate for steady readings on most oil bearing work, though it might be marginal for air bearing work where total gaps may be only 10 $\mu$m. When we come to vibration, 1 $\mu$m is inadequate at high frequencies since a typical vibration level on, say, a large gearbox might be 0.1 *g* which at 1 kHz is only 0.025 $\mu$m. However, this level of vibration would probably be detected on a wave analyser, even though it was not visible on the time record and had an amplitude of only 0.2 mV.

## 3.6 Doppler types

These are non-contact relative velocity devices which measure velocity by the frequency shift of a radio wave caused by movement

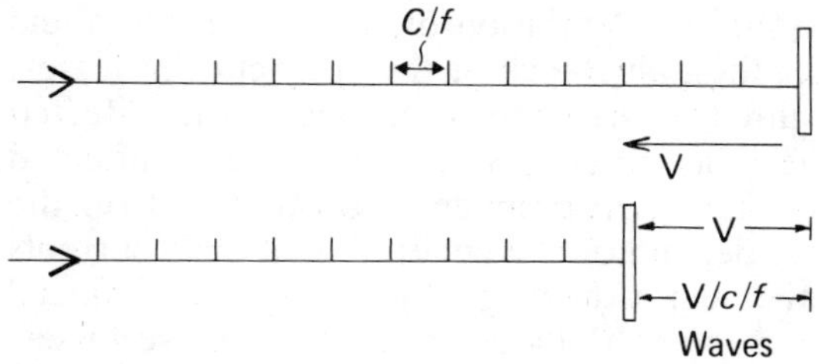


**Figure 3.6** Doppler frequency change principle

of a reflecting target. The speed of an electromagnetic wave, whether a light or radio wave is $c = 3.1\,E\,8$ m/s, so if the transmitted frequency is $f$, the wavelength will be $c/f$. A target speed of $V$ m/s, see Figure 3.6, produces $f + fV/c$ waves each second as an extra distance $V$ is covered, so the frequency shift is $V/c$ of the sending frequency. Although this is a small fraction, radar frequencies are high and so 1 m/s with 10-mm wavelength gives a frequency shift of 100 Hz.

This is the basis of radar automobile speed checks and simple systems will work well at steady speeds, but a system that uses a frequency difference of 100 Hz to detect velocity cannot be expected to observe such high frequencies as, say, 1 kHz. To achieve detection of vibration the returning frequency-modulated signal is compared not with the transmitted signal, but with a derived signal which is perhaps 10 kHz lower. The 'extra' 10 kHz then allows more information to be derived and so a wider range of frequency modulation can be detected [1]. This approach is useful for specialist applications such as observing blade vibrations on a rotating fan, but it is relatively cumbersome and expensive.

It is theoretically possible to use a Doppler frequency shift approach with sound waves. As might be expected a sound wave with a wavelength of 10 mm will give the same frequency shift as a 10-mm radio wave, i.e. 100 Hz per m/s. This system is not used in practice but would be very sensitive to temperature variations. Laser Doppler systems work well, though at much higher frequencies than radar waves, and are the basis of sophisticated replacements for gyroscopic methods of detecting rotation.

## 3.7 Interferometer methods

Measurement of vibration using a laser interferometer can be considered almost as a by-product of laser calibration techniques for

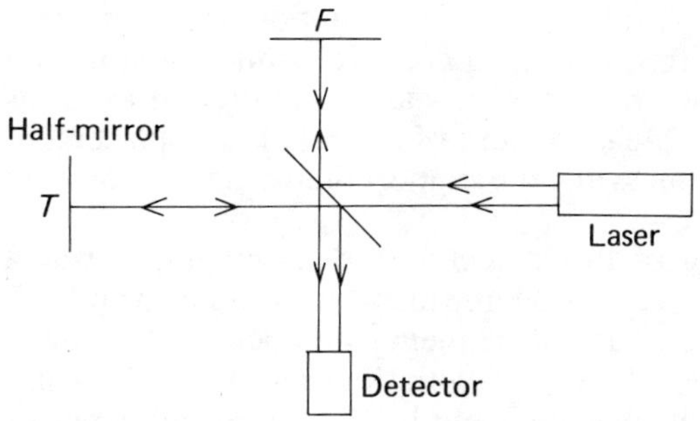


**Figure 3.7** Interferometer principle: *F* is fixed and movement of the target *T* alters the relative phase of the two halves of the beam arriving at the detector

machines. The principle used is that of the standard Michelson interferometer, as illustrated in Figure 3.7. Variation of the target distance doubles the variation in light path and the two beams, one reflected from the target *T* and the other from the fixed mirror *F* will successively reinforce or cancel to give a cycle of light variation (a fringe) at the detector for each half-wavelength movement of the target. The standard He–Ne laser has a wavelength of about 6300 Å units so a fringe occurs every 0.316 $\mu$m.

Refinements to the basic system are required to give not only the main sine wave variation of light intensity at the detector, but also an additional variation at 90° phase, i.e. a cosine to enable the system to determine direction of movement and to subdivide fringes if necessary. The target mirror used is not a simple mirror but a roof prism that will always reflect light exactly parallel to the incoming direction regardless of variation in alignment.

The resulting systems can be used statically or dynamically and are capable of measuring position accurately at velocities of the order of 0.1 m/s.

Fundamental calibration of accelerometers is carried out by laser interferometer by the relatively simple method of counting the total number of fringes in a second and using this to give the total distance travelled [2]. It is best performed at relatively low frequency and large amplitude so that fractions of a fringe at the end of the stroke have relatively little effect.

## 3.8 Seismic systems

The objective in this class of transducers is to measure absolute vibration. The method uses a mass supported on a soft spring to give a low natural frequency. Vibration of the support at frequencies well above the natural frequency will not give significant vibration of the suspended mass, which thus acts as a 'fixed' point as far as vibration is concerned. Measurement of the relative displacement or velocity between the mass and the support then gives the absolute vibration of the support.

The traditional form of this detector, the seismograph, was a rather delicate fixed device with a large mass on a lever arm and soft springs arranged to give a natural frequency well below 1 Hz; it was by no means portable. A typical vibration seismic pickup has stronger springs to give robustness and has a natural frequency of about 10 Hz. The suspended mass may be a magnet and the relative velocity is then easily deduced from the voltage induced in coils mounted on the supporting case. It is possible to have displacement transducers instead of using the moving coil principle, but as this entails higher costs and less sensitive high frequency performance it is not popular.

Limitations on such systems are that the lower end of the frequency range is determined by the natural frequency, and the upper end by contact resonances between the case and the structure under test. Some versions weighed as much as 1 kg and so were limited by contact resonance to about 300 Hz; the limit could be increased by bolting the pickup down but this removed the inherent flexibility of the system.

At the lower end a subtle choice of damping ratio allows these pickups to be used well down towards their natural frequency and still give valid results. The relative vibration between base and mass in a seismic system was given in Section 1.2 and is

$$\mathbf{z} = \mathbf{y}\left(\frac{\omega}{\omega_n}\right)^2 \left\{\left[1-\left(\frac{\omega}{\omega_n}\right)^2\right]^2 + \left[2c\frac{\omega}{\omega_n}\right]^2\right\}^{-1/2}$$

or

$$\mathbf{z} = \mathbf{y}\left\{\left[1-\left(\frac{\omega_n}{\omega}\right)^2\right]^2 + 2c\left[\frac{\omega_n}{\omega}\right]^2\right\}^{-1/2}$$

i.e.

$$\mathbf{y}\left\{1+\left(\frac{\omega_n}{\omega}\right)^4 + (4c^2-2)\left(\frac{\omega_n}{\omega}\right)^2\right\}^{-1/2}$$

Thus if $c^2 = 0.5$, i.e. $c = 0.7$, the second order terms cancel and with $\omega_n/\omega$ of 0.6 the fourth order term gives only 1 dB error, **z** approximates to **y** so the amplitude read is correct but there is some phase distortion.

Use of this type of seismic pickup has decreased as the performance of accelerometers has improved, but in the particular case of low frequency torsional work this method is still used because accelerometers would not be satisfactory at very low frequencies. The application is the measurement of torsional vibration of large slow-moving tables such as hobbing tables and a large moment of inertia is mounted on a crossed leaf spring suspension to give a frequency that can be as low as 0.1 Hz. Non-contact displacement transducers measure the relative movement between inertia and table [3].

## 3.9 Accelerometers

Accelerometers now dominate vibration measurement for all cases other than shaft vibrations. Unlike seismic devices they work well below the natural frequency of a small mass supported on a force transducer. The force required to accelerate the mass is proportional to acceleration until the resonance is approached.

Any form of force measurement would be possible but although strain gauge and piezoresistive systems are used, the majority are piezoelectric. A crystal of quartz or a material such as barium titanate, when stressed, will produce a charge proportional to stress. The crystal may either be used in direct compression, as sketched in Figure 3.8(*a*), or in shear(*b*); compressive types tend to be cheaper, more robust and of higher frequency, whereas shear types show less cross-axis and temperature sensitivity.

A large range of accelerometers is now available; at the high frequency end, natural frequencies are above 100 kHz and mass is less than 3 g but sensitivity is below 0.1 pico coulombs per *g*. At the low frequency end the first stage of electronics is built into a transducer that weighs 500 g, but will work down to 0.2 Hz and has a sensitivity of 10 000 pC per *g*. A normal compromise specification

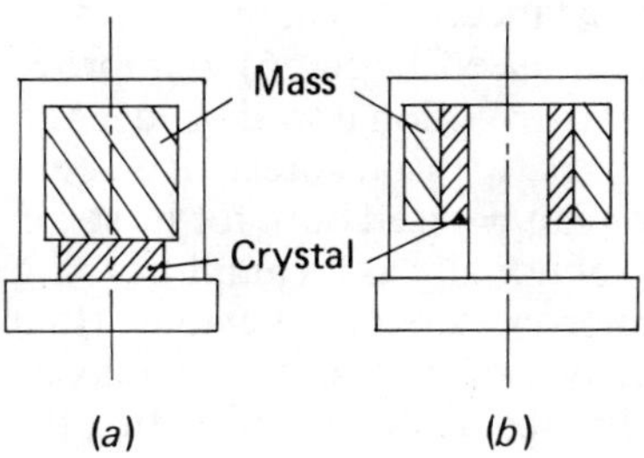


**Figure 3.8** Compression and shear versions of accelerometers

for general purpose vibration work will weigh 20 g, have a useful range up to 10 kHz and give about 30 pC per *g*.

The basic criteria of linearity, discrimination and frequency range are very well satisfied by accelerometers. Electrical matching involves the use of charge amplifiers but this has been made easy with the advent of field effect transistor amplifiers. Mechanically, their low mass makes them suitable for nearly all purposes except light panels, such as in car bodywork or thin plastic components, and their frequency range is good.

Disadvantages stem from a sensitivity to temperature which can give low frequency drift and problems of calibration due to the inability to calibrate statically. A fundamental problem can arise from the nature of the signal because it is acceleration that is measured; in typical machinery there is a high level of high frequency vibration which may dominate the signal of interest. A typical case might involve 5 $\mu$m vibration at a gear tooth frequency of 2 kHz. The high frequency acceleration of nearly *g* will dominate the $2 \times 10^{-3}\,g$ at shaft frequency. The generation of charge instead of current or voltage used to be a disadvantage and the output of the accelerometer is extremely low but modern electronics can cope with these problems. If 1 *g* acceleration is applied to a typical 30 pC/*g* accelerometer with a self-capacitance of 1000 pF the voltage generated is 30 mV and the energy is $0.5\,CV^2$ which is only $5 \times 10^{-13}$ J. Accelerometers can detect below $10^{-3}\,g$ with a corresponding energy level of $10^{-20}$ J or a power level of the order of $10^{-18}$ W at low frequencies.

As the 'spring' in the system is a solid crystal, internal damping is low but ringing at resonance frequency does not usually occur because energy is lost by radiation into the structure on which the accelerometer is mounted.

## 3.10 Resistive accelerometers

These, like the charge type of accelerometer, work below resonance but respond to force by changing resistance. The change of resistance can be detected by a bridge circuit (see Chapter 6) to improve sensitivity. Two types are used, either conventional strain gauges on steel springs or cantilever members, or with piezoresistive semiconductors where the semiconductor can be used both for its strain sensitivity properties and to act conventionally as an amplifier.

Resistive and piezoresistive accelerometers, unlike those that generate charge, will respond to steady acceleration and so may be calibrated very easily by turning them upside down and noting the corresponding 2 *g* static change. This ability is useful for calibration but can be a nuisance if an accelerometer is moved round an object such as a gearbox since the steady level of output changes with orientation.

Choosing between charge and resistive types is difficult. The resistive is easily calibrated but the charge type is preferred when stability, linearity or sensitivity is extremely important. This is principally due to the use of balanced crystal shear designs. Alternatively, the strain gauge type of accelerometer is relatively good at low frequencies, even to DC and does not suffer from drift but has limited high frequency performance. Most installations use charge accelerometers because their range is large and the electronics is much simpler than with strain gauge bridge circuits, which need balancing.

## 3.11 Accelerometer calibration

Calibration presents problems because although it is possible statically with some electronics, it is marginally satisfactory for acceleration by turning the accelerometer upside down, drift problems are troublesome and subsequent integrals to velocity or displacement cannot be checked this way.

The most fundamental method is the use of laser interferometer counting techniques mentioned in Section 3.7, but this can only be used under laboratory conditions and since, essentially, a distance is checked against an acceleration, the waveform must be a good sinusoid with very low harmonic content. The most convenient method is to buy a calibrator and to have faith that when it says $\pm 1\,g$ the device is giving exactly that. One type of calibrator works vertically and relies on an audible rattle to detect loss of contact of a loose ball to indicate 1 *g* but this test is open to debate since the borderline is not easily located.

Back-to-back testing against a known accelerometer is straightforward if a standard accelerometer with its electronics is available, but this is rare. The alternative is to use a vibrator with a fairly long stroke at low frequency so that the amplitude of vibration is relatively large and to measure displacement with a travelling microscope or, surprisingly accurately, with a 6 inch rule. A traditional method is to draw a small target, as shown in Figure 3.9(*a*), with diagonal lines at angle arctan 0.5; when vibrated the target appears as in Figure 3.9(*b*) and the amplitude can be set to make the central rhombus touch the vertical lines with an accuracy of about 2%. The main problem with this method, as with the laser method, is to

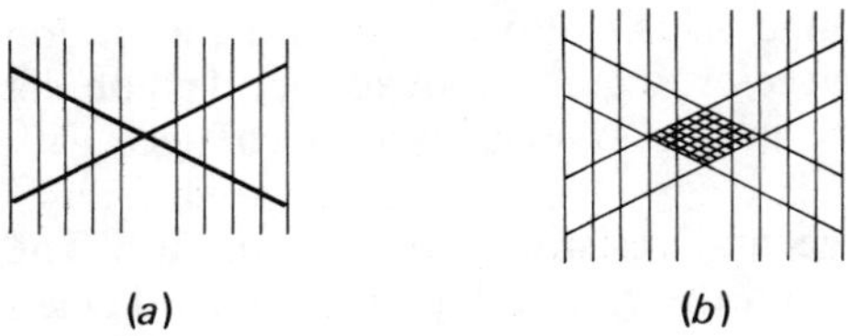

Figure 3.9 Diagonal lines flag method of calibration

obtain a waveform with negligible harmonic content. This is achieved by attaching a large mass to the vibrator so that inertia effects rather than spring effects dominate, and the resulting waveform is then satisfactory for acceleration and, due to the effects of the integration, is very good for displacement. However, if the set up is used to calibrate output directly in the displacement form, purity of waveform is not so important.

Calibration of accelerometers for use with shock testing is rather more difficult to extrapolate from a test at 1 $g$. In this case a good check on calibration can be obtained by integrating the area under the acceleration pulse in a test and comparing it with the overall change of velocity in the impact; as the velocities before and after are fairly steady they are relatively easy to measure using passage time between photocells.

**Table 3.1 Common transducers**

| *Problem*<br>*Amplitude, frequency, relative or absolute* | *Pickup* | *Comments* |
|---|---|---|
| Medium, low, relative | LVDT | Electronic dial gauge. Less suitable for high frequencies<br>Contact |
| Medium, low, relative | Resistance slider | Good phase characteristic<br>Contact |
| Low or high, high, absolute | Accelerometer (small) | Most general usage<br>HF limit > 50 kHz |
| Low or high, low absolute | Accelerometer (int. electronics) | Frequency limit of large accelerometers is about 0.1 Hz |
| Very low, low or high, relative | Eddy current, non-contact | Especially useful for rotating shafts and bearings up to 2 mm and 10 kHz |
| Very high, low, relative | Grating/ inductosyn/ interferometer | Basis of measuring machines and NC machine tools (see Chapter 5) |

## 3.12 Choice

When presented with a fresh vibration measurement problem there is no absolute or 'right' answer for choice of transducer and much will depend on what is already available. The general areas of application for the common transducers can be summarised easily, as in Table 3.1.

It is, of course, possible to make a relative measurement using two accelerometers and this method is satisfactory provided the movement is not a small difference of two large quantities. Correspondingly an 'absolute' measurement can be made if there is a fixed base available nearby or if a stationary point can be synthesised using a very low frequency suspended mass.

When the problem does not fall into these areas special techniques such as the Doppler radar or laser approach may be needed.

Chapter 4

# Measurement of sound

## 4.1 Introduction

It should first be commented that 'noise' is a confusing term because it may be used in two different senses, to mean either airborne sound waves or background electrical vibrations that are not due to the vibration being measured. It will simplify life to refer to 'noise' purely in an electrical context and to specify 'sound' for airborne pressure waves.

Measurement of sound might seem out of place in a text on vibration measurement but the sound from a device is a measure of some form of averaged vibration over the whole surface of the device, and since it is eventually sound that irritates or betrays the customer it is sometimes advisable to measure it. Alternatively, a microphone held near a sound source can be used as a crude form of vibration measurement. In some cases, as in the machine guard illustrated in Figure 4.1, it is essential to scan with a microphone to see whether the sound is emitted from the guard itself or is leaking through the apertures.

The sound from a vibrating system travels relatively slowly through air, at about 300 m/s. At 1 kHz the corresponding wavelength is 300 mm and thus if the sound from two points on an engine

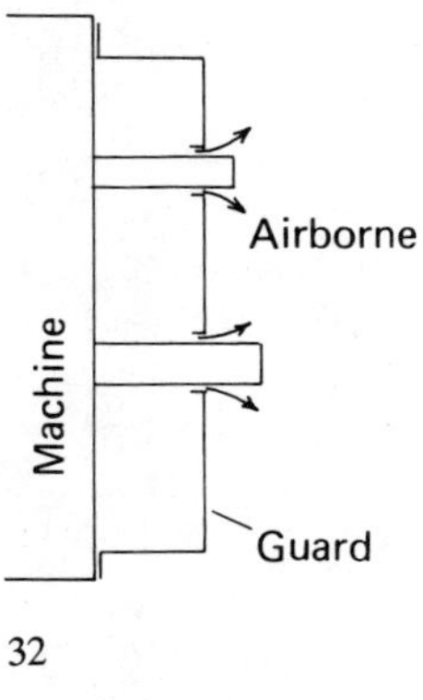


**Figure 4.1** Noise generation may be either by leaking through unsealed gaps or by panels vibrating

block is emitted simultaneously, but the distances to the microphone differ by 150 mm, the sound effects will tend to cancel instead of add. Similarly, a single shock pulse, appearing simultaneously at two panels, will be heard as two distinct pulses at the microphone unless distances are equalised.

Such delays greatly complicate the observed sound signal and make any phase information rather distorted since even the conventional 1 m distance would involve a correction of over 1000° in phase angle (at 1 kHz).

Despite these complications, the use of a sound measurement is extremely useful for problems such as engine noise since it quickly supplies information on the type of waveform (steady or impulsive), the areas from which the trouble comes, and whether the trouble occurs in specific, well-defined parts of the firing cycle or all the time. This type of information will assist the placing of accelerometers and identification of trouble areas.

In cases where the trouble source is suspected, e.g. a timing cover on an engine, it is worth while placing a microphone square to the cover so that waves from all parts of the cover arrive at the same time and, of course, there should be a minimum number of reflecting surfaces around.

## 4.2 Scales

Sound is an air vibration in the form of an alternating compression and rarefaction with no shearing effects, and could be specified either in terms of the movement of the air or of the variation in air pressure. The latter is more useful as well as being easier to measure and the basis for pressure variations has been chosen as $2 \times 10^{-5}$ N/m$^2$, intended to correspond to the lowest level of sound that can be heard. The figure is given in rms and so peak values are $2^{1/2}$ higher for a sinusoid.

The scale is logarithmic on the usual decibel basis so that 20 ln (pressure ratio) is the quoted figure; 124 dB corresponds to 40 N/m$^2$ and so is still less than 0.001 atmospheres. This pressure variation can be achieved mechanically fairly accurately and so is often used for calibration purposes. For most engineering work specification of dB in particular octave frequency bands is sufficient. Much research work has gone into the characteristics of human hearing and so there are systems for 'adjusting' the raw pressure variation information to match human sensitivites or damage limitations. The simplest way is to apply a weighting curve to the measured pressure level output in the hope that this is a good approximation to human response. Curve *A* of the standard curves is usually used and is a reasonable average, though at very high or low sound levels it is less relevant (see Figure 4.2).

If more detailed frequency and amplitude 'adjustments' must be applied then the raw data of the sound pressure level variations with frequency may be corrected to 'phons' which are identical to dB sound pressure level at 1000 Hz, the reference frequency, and follow experimentally determined curves that reflect the variation of human sensitivity with frequency and amplitude to 'equal loudness'. Typically, at 50 Hz a sound of equal 'loudness' has 15 dB higher pressure level than at 1000 Hz.

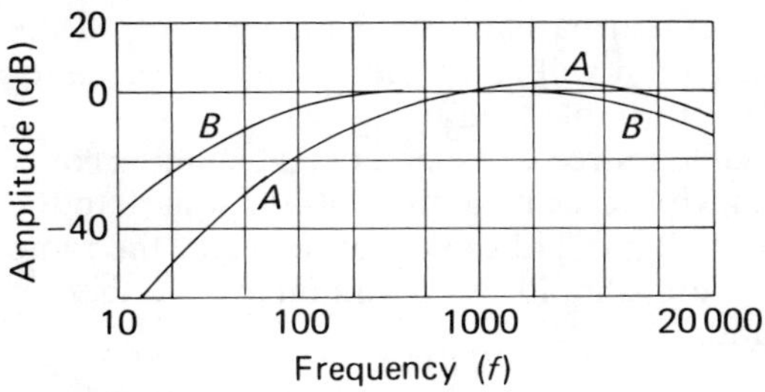


**Figure 4.2** Standard sensitivity weighting curves

The 'harmfulness' of the noise involves conversion to noise rating (NR) values by means of curves, again experimentally determined, which rate noise damage abilities, usually in octave or third-octave bands, against frequency. The comparison is again with 1000 Hz so that a single tone at 1000 Hz has the same NR value as its sound pressure level. The curves differ from the *A*-weighting and 'phon' curves in that it is difficult to cause damage at low frequencies so, at 50 Hz, a 100 dB sound pressure level has a NR of 70, but at high frequencies (in particular at about 4 kHz) ears are more vulnerable than at 1000 Hz. Research is still proceeding on human reactions to sound and so standards are updated steadily; ISO specifications and recommendations R532, R1760, R1996 and R1999 are relevant. See refs. [1, 2] for a range of literature on the subject.

For vibration work there is no point in converting such pressure levels into dB(*A*), phons or NR. Absolute values of noise are not usually important since it is frequencies, regularity and customer irritation that matter and thus the primary requirement is identification of the causes.

## 4.3 Measurement

Sound measurement involves measuring variations in air pressure which are small and at high frequencies. The two standard methods are the use of piezoelectric crystals, as used in accelerometers, and the capacitance variation microphone. As the former is simple and robust but is more affected by ambient conditions, the capacitance type is favoured for accurate work. Figure 4.3 indicates the principle on which this works.

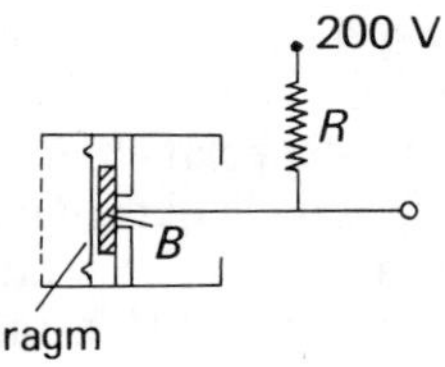


**Figure 4.3** Principle of capacity microphone

Air pressure fluctuations incident on the very thin flexible diaphragm produce a deflection that is proportional to pressure and vary the capacitance between the diaphragm, at earth potential, and the central fixed plate B; the gap is small, normally 22 $\mu$m. Detection of the capacity variation is achieved by an unusual method that involves generating a 200 V DC supply from the 1.5 V batteries that power the amplifier, and connecting it to the capacitor plate *B* via a very large resistor R. This 'traps' a charge in the capacity and the time constant of the resistance *R* and capacitance is of the order of 0.1 s so that, dynamically, charge cannot leak out. Since $q = CV$ and $q$ is fixed, variation of $C$ alters $V$ and this variation is relatively large since a 0.1% variation in gap, and hence $C$, will give 0.1% of 200 V; 0.02 $\mu$m gives 200 mV.

The equipment is very linear, has a frequency range of 20 Hz–20 kHz, and calibration can be checked easily. Problems in sound measurement rarely stem from the equipment but are usually due to reflection of sound waves from neighbouring obstacles.

## 4.4 Effects of sound fields

The most common problem in sound measurement is the effect of walls and ceilings in distorting the pattern of sound level. In a free field, with a source of relatively small dimension, the sound pressure level will follow an inverse square law and so the total sound power emitted by the machine can be deduced relatively easily.

Reflecting surfaces prevent simple measurements from giving worth while results unless a reverberation chamber is used to give so much sound reflection in random directions that the sound level in the room is approximately constant, regardless of position. The machine under test is then removed and a calibrated noise source is substituted.

An anechoic chamber is an extreme case in which there is no reflection and completely free-field conditions apply, but rarely can tests on high power devices be carried out in such a chamber and use of a chamber for routine testing of low power devices is too expensive. In the particular case of noise inside a vehicle body there is, of course, no point in having the vehicle in a special chamber.

Most tests, such as machine tool checks or vehicle external noise checks, are specified to be carried out on a flat, free surface. A concrete floor gives repeatable conditions since the reflection from the floor is high and consistant and the power level at a distant point is roughly double the level for a similar source in an anechoic chamber. Unfortunately, although it is easy to arrange a free space for vehicle testing, it is difficult to obtain a sufficiently large space inside a factory.

In a typical restricted space with some reflecting surfaces, the technique of substituting a calibrated sound source can still be used. However, as each frequency will be reflected differently the results for each frequency band must be processed separately in order to find the relationship between sound power and observed pressure levels. With reflecting surfaces there is the possibility of standing waves and thus large variations in sound pressure level with position of both source and microphone. Inside a vehicle body, reflection is very important and, combined with several sources, can give sound pressure levels at certain points higher than the pressure levels at the panels that are generating the sound; pressure levels can then be very deceptive in locating sources.

To overcome some of these problems, techniques of sound intensity measurement have been developed recently.

## 4.5 Sound intensity

The fundamental problem with simple measurement of sound pressure levels arises from the fact that it is a scalar quantity which is being measured and if a high level of sound pressure is measured we do not know whether it is due to a powerful source or to a weak

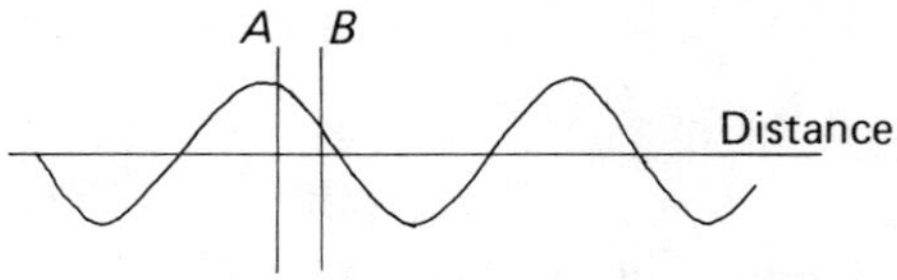


**Figure 4.4** Sound intensity measurement. The local velocity is deduced from the gradient between $A$ and $B$

source with a very reverberant reflecting field. The simple answer is to have a non-reflecting field so that we know the shape of the sound pressure field (as it follows an inverse square law) and we can then deduce the strength of the source that is supplying the field.

An alternative answer, the only possible one in some circumstances, is not to measure the pressure level at a point but, rather, the energy transmission. This 'sound intensity' is more appropriately termed an energy flux and has units of W/m$^2$. The basic theory of the measurement is described clearly in a publication by Bruel and Kjaer [3]. Since we require an energy flow (per unit area) this can be obtained instantaneously as the product of a force and a velocity. Referring to Figure 4.4, the force on the air at position $A$ is easily determined since it is given by the sound pressure level at position $A$ and can be measured with a normal microphone. Any steady pressure, such as atmospheric pressure, can be ignored since we are only interested in vibrations. The velocity of the air could be determined by laser Doppler or hot wire methods but it is easier to mount a second, identical microphone at a near point $B$ to give the pressure gradient between $A$ and $B$, assumed linear. This pressure gradient gives the acceleration of the air, and integration of the acceleration gives the velocity of the air. Product of force and velocity in power flow can then be integrated to give total power flow.

In the case of a free field with no reflections the equation of a plane propagating wave is $p = p_o \sin(\omega t - 2\pi x/\lambda)$ where $x$ is the direction of travel and $\lambda$ is the wavelength.

$$\text{Pressure gradient } \frac{\mathrm{d}p}{\mathrm{d}x} = p_o \cos(\omega t - 2\pi x/\lambda) \cdot (-2\pi/\lambda).$$

The instantaneous acceleration is pressure gradient/mass and so is

$$\frac{p_o(-2\pi)}{\lambda \varrho} \cdot \cos(\omega t - 2\pi x/\lambda),$$

where $\varrho$ is density. To obtain the velocity, integrating with respect to time gives

$$\mathbf{v} = \frac{-2\pi p_o}{\lambda\varrho} \cdot \sin(\omega t - 2\pi x/\lambda) \cdot \frac{1}{\omega}.$$

The power is the pressure–velocity product, thus

$$\text{Energy flux} = \frac{-2\pi p_o^2}{\lambda\omega\varrho} \cdot \sin^2(\omega t - 2\pi x/\lambda),$$

which has an average value of $p_{rms}^2/\lambda f\varrho$ since frequency $f = \omega/2\pi$ and $\lambda f$ is the velocity of sound $c$ (in air typically 330 m/s).

Actual power flows are small; at 100 dB sound pressure level $p_{rms}$ is 2 N/m$^2$, so with $c = 330$ m/s and air density $\varrho$ of about 1.4 kg/m$^3$ the energy flow is $2^2/1.4 \times 330$, which is only 0.01 W/m$^2$.

In a standing wave field, for example an organ pipe, there is no net flow of energy. The pressure varies along the pipe but the pressure variation is all in phase with the equation:

$$p = p_o \sin \omega t \sin 2\pi x/\lambda.$$

In this case the pressure gradient is still in phase with the pressure and so the velocity, after integration with respect to time, is 90° out of phase with pressure and the product integrates to zero.

There are various approaches to the calculation of the sound intensity from the two microphone readings: either operate directly on the time signals or first carry out a Fourier analysis (with the necessary phase information) and then multiply. Choice is governed by the available equipment and its speed and also whether the information is required as narrow band or as third octave information. As it is pressure gradient that is critical, very accurate matching of the two microphones is required, and the distance apart must be small compared with the wavelength.

Chapter 5

# Digital measurement systems

## 5.1 Introduction

The systems so far described are generally intended for measurements of relatively small amplitudes and high frequencies, but are hopeless for the type of measurement required for numerically controlled machine tools or for coordinate measuring machines (CMMs). Achievement of 0.1% accuracy is possible with a normal vibration transducer but an accuracy of 1 $\mu$m in a meter is required for NC or CMM, and both static and dynamic accuracy are needed. The only system so far mentioned that can achieve this is the laser interferometer and this approach is used for extremely accurate CMM reference machines but is too expensive for general use.

The solution that was found is a hybrid system which is part digital and part analogue, analogous to optical scale reading instruments in which there is an accurate but relatively coarse main scale and a separate system for interpolation between the scale divisions.

## 5.2 Fringe counting systems

Optical systems use a grating consisting of precisely spaced lines on glass at a pitch of typically 100–250 $\mu$m. The line thickness is half the pitch, and the grating, which is usually stationary, is viewed by an index that is a short section of similar grating carefully aligned with the main grating. The effect is as shown in Figure 5.1; a collimated beam of light will be transmitted (50%) when the lines are aligned, while little will be transmitted when index lines coincide with grating spaces. Movement of the index will generate a roughly triangular variation in light and each complete pitch or fringe corresponds to a cycle of light intensity transmitted; a photocell then converts this to voltage.

Alternatively, a reflective steel tape is used, avoiding some of the problems of fragility and thermal expansion of glass, but the principle is the same.

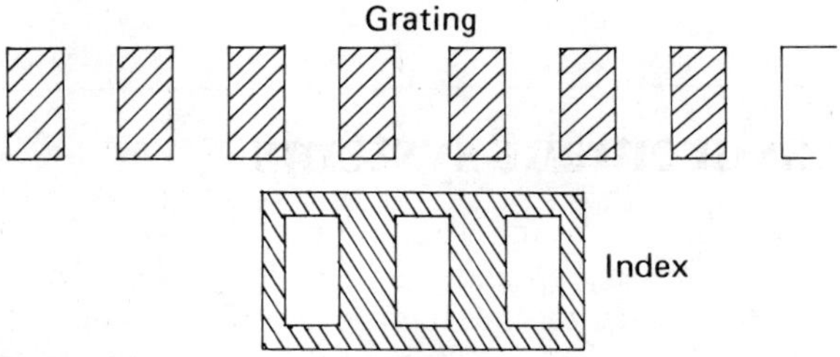


**Figure 5.1** Relative position of index and grating for maximum light transmission. Usually the index is stationary to simplify wiring

Such a system will easily count complete fringes for unidirectional movement, but it cannot determine direction of motion or interpolate between fringes. Figure 5.2 attempts to show a system used for interpolation: four distinct index gratings, each with its detection photocell, are used and are mounted so that they are optically one-quarter of a pitch apart. The four photocells are then scanned cyclically at a frequency of several kHz. If the composite index is in position *A* the cells will read medium, high, medium, low, etc., approximating to a sine wave. In position *B* the output will be high, medium, low, medium, etc., and hence a cosine wave 90° in advance of the sine wave from *A*.

The phase of the output wave relative to the scanning sequence thus gives a measure of the position of the index within a fringe. Since phase measurement of signals is straightforward, interpolation can be achieved to, say, 1.5 degrees of phase, which is the micron required even with a grating of 250 $\mu$m pitch. Once

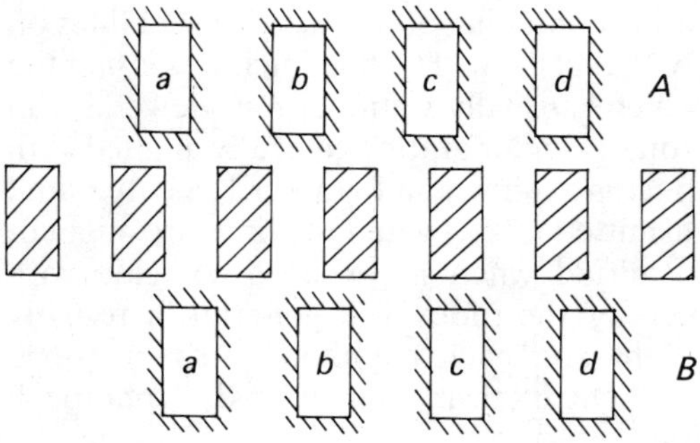


**Figure 5.2** Principle of interpolation. *a*, *b*, *c* and *d* are four separate index sections which are scanned electronically. Relative movement from position *A* to position *B* advances the phase of the output by 90°

interpolation has been achieved it is relatively easy to determine direction of movement and hence whether main fringe transitions should be added or subtracted.

Accuracies are remarkably high and errors on CMMs are more likely to be due to slideway straightness or deflection problems than to the grating system. There is no inherent limit on the accuracy of gratings; masters are made by optical methods involving photographic reduction of large originals and thus an inherent increase in absolute accuracy. Short sections can then be assembled to breed large, accurate machines to produce more accurate masters for reduction. The effects of residual errors on gratings are then reduced by using an index that is relatively long axially and so averages over perhaps 100 lines on the grating.

The same principles are used with rotary gratings, though it is common to use two index assemblies mounted 180° apart to cancel out eccentricity and even-harmonic errors; alternatively an optical system can be used on the two sides of the grating, as shown in Figure 5.3, and the image of one side is superposed on the other side. With radial gratings it is more dificult to achieve high accuracy since photographic reduction techniques do not help but 0.1 seconds of arc, corresponding to 1 $\mu$m at a 2 m radius, can be achieved.

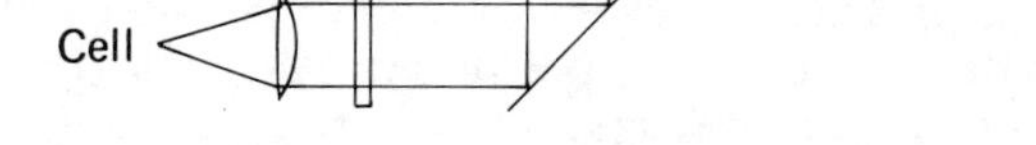


**Figure 5.3** Use of circular grating acting as its own index

## 5.3 Laser and electrical systems

Laser fringes, derived from a Michelson interferometer can be subdivided but there is little requirement as the wavelength of He–Ne lasers is about 6300 Å and so a fringe corresponds to 0.315 $\mu$m. The corresponding disadvantages are a velocity limit since 0.1 m/s gives 300 kHz pulse frequency and correction for temperature, pressure and humidity is necessary.

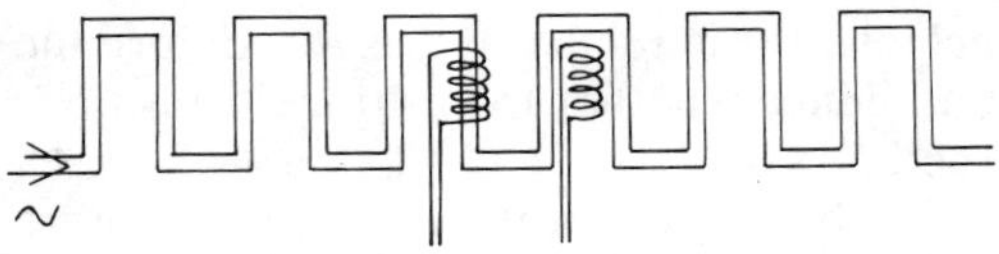

**Figure 5.4** Principle of electrical system. When a coil is directly over a conductor there will be maximum voltage and mid-way between will be zero

Electrical systems are based on a printed circuit of high accuracy photographically etched conductor in the form of a square wave, as illustrated in Figure 5.4. If an alternating current is passed along the 'wire' and a small coil is placed over the wire a voltage will be induced in the coil, changing direction as it passes each conductor and with zero voltage mid-way between conductors. A second coil, placed one-quarter of a pitch later will give the additional quadrature information necessary to determine direction or to interpolate. The system requires only two coils instead of four photocells because each coil gives two items of information – amplitude and phase – rather than amplitude alone (i.e. light intensity).

## 5.4 T.E. measurement

The systems described above are predominantly used for static measurement or for machine tool control. A slightly different approach is used to check the smoothness of gear drives for control of gear generated noise.

Transmission error (TE) is defined as the difference between the actual position of an output gear and the position in which it should be if the drive were rigid and perfectly smooth. One radial grating is attached to the input gear and one to the output gear and the system is run at low speed. The sine wave from the input gear is turned into a string of pulses, typically spaced at exactly 1 minute of arc, i.e. 21 600 pulses per revolution. This string of pulses is frequency multiplied and divided electronically to generate the string of pulses expected at the output. Any variation of the phase between expected and actual output pulses gives the variation of position of the output gear. Figure 5.5 is a block diagram of the system.

Accuracy is high, of the order of 1 second of arc or better, and the test shows up pitching errors and once per tooth errors which can be due to helix or profile errors. As normally used in inspection, speeds are less than 10 rpm but it is possible to use small robust encoders

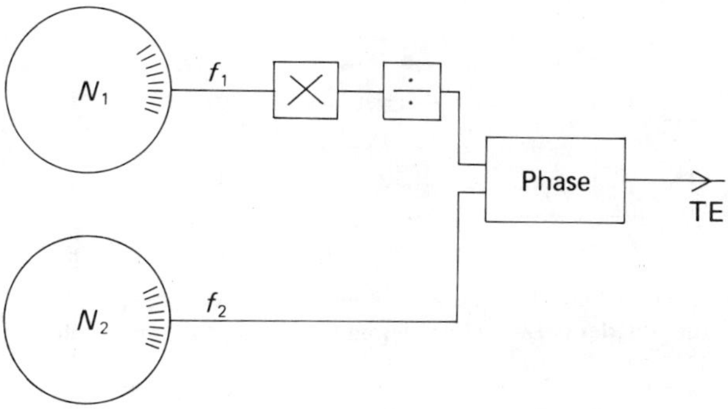


**Figure 5.5** Principle of error measurement. Pulses from the encoders are multiplied and divided in frequency to give the same average frequency, and then compared for phase

that are slightly less accurate but can run at several thousand rpm and give useful information up to about 1200 Hz.

An alternative method of processing the information for inspection is to use a computer to carry out the sampling and analysis [1], but it is difficult to obtain a computer that is sufficiently fast to operate at full speed for development work. The normal approach is to subdivide the intervals so that the input encoder is providing around 100 pulses per line, i.e. minutes of arc, and then use the pulses from the output encoder to batch count the input pulses and compare with the expected 100 $N_2/N_1$ pulses.

## 5.5 Torsional vibration

A further variant on TE measurement is to take the string of pulses from an encoder and to compare phase with a steady string of pulses from an oscillator. This comparison will give the fluctuations in position about the mean speed and so will respond as a torsional vibration transducer.

An oscillator under manual control would be difficult to set and any variation in test speed would give drift off scale, thus the technique is to use a phase locked loop. The circuit is shown in Figure 5.6; the incoming (variable) string of pulses is used to switch the flip-flop up and the pulses from the voltage controlled oscillator (VCO) switch it down. The resulting variable mark-space ratio square wave is smoothed and integrated to drive the oscillator which will then follow the frequency of the incoming train. If the time

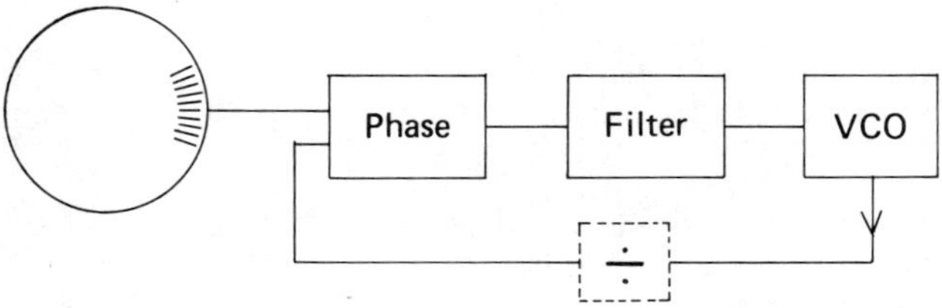


**Figure 5.6** Principle of phase locked loop used to measure vibrations from an average speed. Another divider between encoder and phase measurement will allow reduction of sensitivity

constants of the circuit are short, the oscillator follows rapidly and, with the extra divider in circuit (dashed) will give a frequency multiplication. When the time constants are long, of the order of 1 second, the oscillator will follow slow speed drifts of the rig but will not follow relatively fast torsional vibrations of the encoder. As with TE measurements, the variation in phase between the two strings of pulses gives the angular displacement.

Analysis of phase-lock loops is fairly straightforward as they behave as second order systems with a single natural frequency and a damping ratio controlled by a series resistor $R_o$; the transition from frequency to phase is an integration process and the circuit characteristics are given by

$$f = KV,$$

$$\phi = \int \delta f \, dt,$$

$$V = \left(\frac{R_o}{R_i} + \frac{1}{sR_i C}\right) V_\phi,$$

$K$ in (rad/s)/V is the characteristic of the VCO, $\phi$ will increase 1 radian each second if the frequency difference is 1 rad/s; $k$ is the V/rad phase from the flip-flop. $R_o$ is in series with integrating capacitor C; $R_i$ is the integrator input resistance.

This gives

$$s^2 + kK\frac{R_o}{R_i} \cdot s + \frac{kK}{R_i C} = 0$$

as the characteristic equation and

$$\omega_n^2 = \frac{kK}{R_i C}$$

and the damping ratio

$$c = \frac{kKR_o}{2\omega_n R_i}.$$

The voltage at entry to the VCO will have some ripple on it from the square wave but this is usually small.

Typical figures for a circuit at about 1.6 kHz might be

| | |
|---|---|
| $K$ | is 2000 (rad/s)/V, |
| $k$ | 0.5 V/radian phase difference, |
| $R_i$ | 1 MΩ, |
| $R_o$ | 100 kΩ, |
| $C$ | 10 micro F, |

giving a natural frequency of 16 Hz and a damping of around 0.5 of critical.

The output (phase) from the circuit, smoothed through a sharp cutoff low pass filter, set to about one-quarter of the pulse frequency, will give the torsional oscillations of the system reliably down to the natural frequency. Two such gratings will give the TE or, somewhat simpler, if a large flywheel can be fitted at input to suppress torsional movement the one transducer will give TE.

As with TE at low speeds it is possible to feed the string of pulses directly into a computer and to analyse the time intervals between pulses to derive the speed variation information. When possible this method has no lower frequency limit and the upper limit is simply governed by the sampling frequency. In practice this method is little used as yet because high data transfer rates are required. A 2 minute grating at 60 rpm requires 10 kHz data rate on a channel and few computers can exceed this speed.

## 5.6 Torsional comparisons

There are many methods of measuring torsional vibrations including:

1. Seismic transducers: motion between inertia and shaft is detected by a displacement transducer and transmitted by sliprings.
2. Rotary encoders.
3. Magnetic track.
4. Tachometer on contacting wheel with rubber tyre.
5. Double tangential accelerometer.
6. Doppler methods.

1 and 2 require a free end on a shaft and 3 has not yet been successful at high speed for vibration work; 4 is very useful for shafts where the vibrating velocity is significant compared with the mean velocity but is limited by the resonance of the friction tyre drive. A modern variant [2] gives an acceleration signal by inducing voltage in a solid conducting plate; 5 is likely to be useful at high frequencies but does not work well at once per revolution frequencies and requires slip rings or telemetry.

Traditionally, most internal combustion engines were used 1 with a torsional mass having a natural frequency of the order of 10 Hz and, since the output was displacement, the system was very useful for relatively low frequencies but less useful for high frequency work.

Although 6 is a useful non-contact method, a good signal to noise ratio requires the vibration velocity to be significant compared with the steady velocity, as with 4.

Chapter 6

# Signal conditioning

## 6.1 Operational amplifiers

The basis of much conditioning of signals ready for recording or analysis is the 'operational amplifier'. This is simply an amplifier with a low frequency gain of over 10 000 and an output that is 180° out of phase with the input, i.e. if the input goes up the output goes down. There is an additional 'non-inverting' input that is usually connected directly to earth.

Amplifiers commonly work between supplies or 'rails' at ± 15 or ± 12 V and there is normally a balance system to allow setting up, such that zero input corresponds to zero output. The ubiquitous 741 amplifier is typical, widely used, and capable of giving an output current of 5 mA. The amplification as a function of frequency is of the form shown in Figure 6.1.

The old-fashioned name for op-amps was the 'virtual earth amplifier ', a name that gives a clearer idea of the function. Figure 6.2 shows the circuit and, provided the amplifier output is within its linear range of ± 10 V, the high gain ensures that the input cannot be more than 1 mV and so is virtually at earth potential, though not connected to earth.

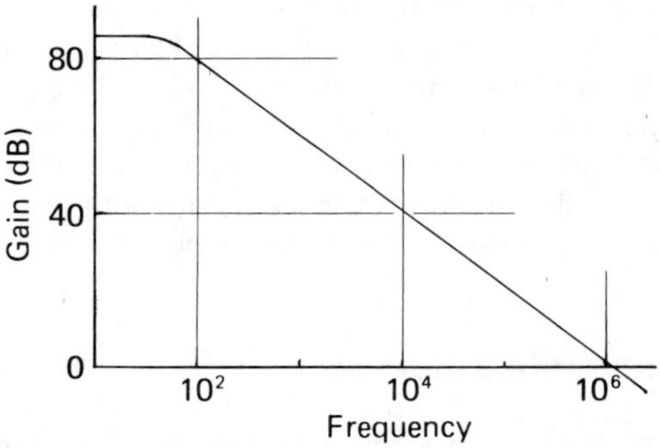


**Figure 6.1** Typical operational amplifier response

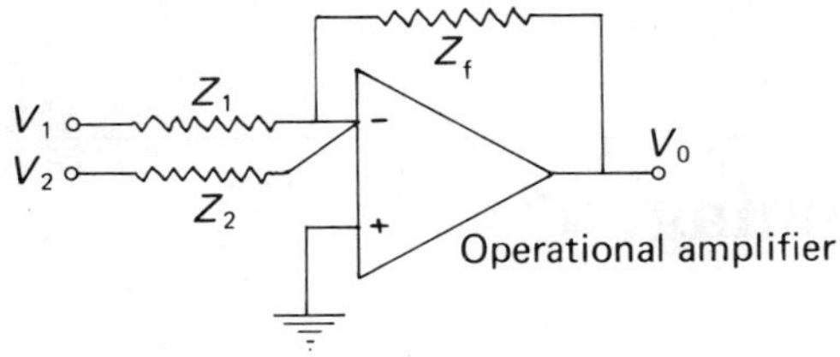


**Figure 6.2** Normal use of operational amplifiers. Impedances may be any combination of resistors and capacitors

Provided the currents into the amplifier input are negligible the current through the input impedances $Z_1$ and $Z_2$ must balance the current through $Z_f$ and so

$$\frac{V_1}{Z_1} + \frac{V_2}{Z_2} + \frac{V_o}{Z_f} = 0$$

Choice of $Z_1$ and $Z_f$ gives all the functions normally required and since only passive and highly linear resistors and capacitors are used the signal will not be distorted.

Limitations on the standard 741 are the frequency response, which is restricted to 10 kHz at full voltage swing, the relatively high input currents, and a low input impedance of the order of $10^6\,\Omega$.

Subsequent developments have given many variants of which the most popular is probably the 3140, an amplifier with field effect transistors on the input, giving an impedance of $10^{11}\,\Omega$ and with improved frequency response. Frequency response is controlled in two ways. Firstly the 'slew rate' specifies the maximum rate at which the output voltage can alter, regardless of input drive; this parameter will be typically 6 V/$\mu$s and thus it is possible to obtain a 20 V swing in little more than 3 $\mu$s so that 200 kHz sinusoidal is achievable. The other parameter, the small signal response, is specified as a gain–frequency product and for the 3140 is 4 MHz so that at 40 kHz the input voltage variation will be 1% of the output voltage swing. By 400 kHz the effect is large, with the 'virtual earth' swinging 10% of the output voltage and so introducing large errors. Any attempt to obtain large amplifications at these high frequencies would be a waste of time and a much higher frequency amplifier must be used, thereby involving attendant stability problems.

## 6.2 Scalers, buffers and integrators

Choice of $Z_1$ and $Z_f$ as simple resistors gives direct scaling of output

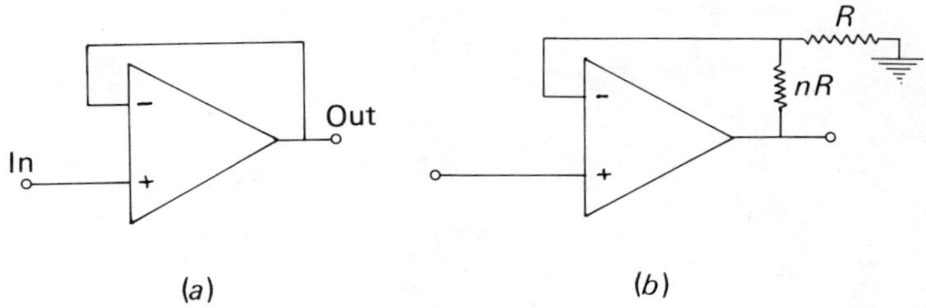


**Figure 6.3** High input impedance circuits: (*a*) gives simple buffering and (*b*) gives a gain of $n + 1$

to input and if, instead of a single input, there are two inputs, the combined currents are balanced and thus the output is the sum of the two inputs, which can be scaled individually. A difference of two signals can be achieved either by a more complex circuit using the 'non-inverting' input or by using another amplifier as a simple 1 : 1 inverter giving only 180° phase shift.

Buffering is achieved since a typical input impedance of 100 kΩ will only drain 0.1 mA from the source but the output can deliver at least 5 mA. In some cases even 0.1 mA may be too high and then the circuit shown in Figure 6.3(*a*) is used with the input connected solely to the non-inverting input of a 3140 with an impedance of $10^{11}$ Ω. The amplifier operates to eliminate any voltage difference between its inputs and so the output follows the input exactly, without input drain. It is of course possible to use (*b*) and obtain a gain in the system without input loading; this type of circuit is used if current must be minimised, as with transmission through resistive slip rings.

A single input through a resistor and a capacitor $C$ as feedback gives a current balance that $V_i/R_i + C\,dV_o/dt = 0$ since the current through a capacitor with $q = CV$ is obtained by differentiating. This gives $V_o$ as the integral of $V_i$ divided by $R_i\,C$, the time constant, and so acceleration can easily be integrated to velocity and then to displacement. The snag with the circuit is that any residual input voltage or imbalance in the amplifier will allow the output to drift until it saturates against the supply voltage rail. A resistor in parallel with the capacitor will reduce amplifier effects but input mean voltages will be amplified by $R_f/R_i$ which may well be 1000; and thus it is sometimes necessary to have an additional blocking capacitor at input, as shown in Figure 6.4. Typical figures for such a circuit for vibration use might then be $R_i = 100$ kΩ, $C_f = 0.01\,\mu$F, $C_i = 1\,\mu$F, $R_f = 10$ MΩ. The time constant of integration is $10^{-3}$ s so that a sine wave input at $10^3$ rad/s would be the same amplitude at output (but 90° out of phase). The blocking capacitor at input and antidrift

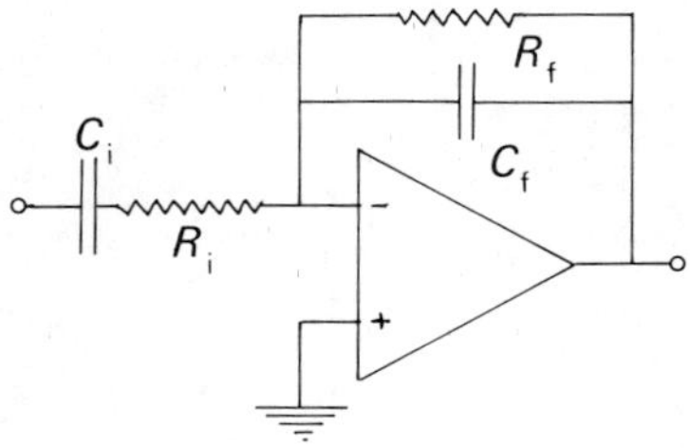


**Figure 6.4** Typical practical integrating circuit; $C_i$ and $R_f$ are added to prevent drift

feedback resistor both give time constants of 0.1 s and thus the circuit will fail to integrate properly as the frequency drops towards 10 rad/s, i.e. 1.6 Hz, but would be satisfactory above 3 Hz.

## 6.3 Filters and differentiators

The circuit of Figure 6.5(*a*) will in theory differentiate an input signal since the current through the input capacitor is proportional to the rate of change in input voltage. Two problems which arise are that high frequency input noise is differentiated and produces large signals, and at high frequencies the input capacitor has a low impedance and may draw excessive current from the previous stage. Both problems are usually solved by placing a small resistor in series with the input capacitor, typically one-tenth of the feedback resistor; this limits the gain to 10. An alternative is to use a complete feedback loop with integration in the return path, as shown in Figure 6.5(*b*). Fortunately, differentiation is rarely required; it is also achieved very simply in a computer by taking the difference between successive samples.

Simple filter circuits can be used in the form of exponential lags to give high frequency cutoff, as illustrated in Figure 6.6(*a*), or

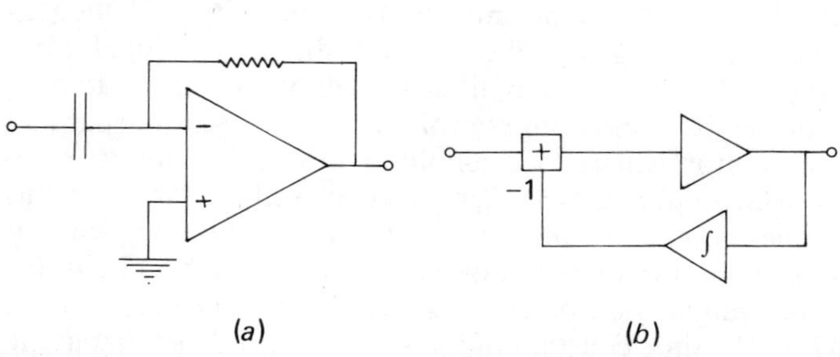


**Figure 6.5** Differentiation. The simple circuit (*a*) poses problems and so needs an extra series input resistor; (*b*) poses drift problems

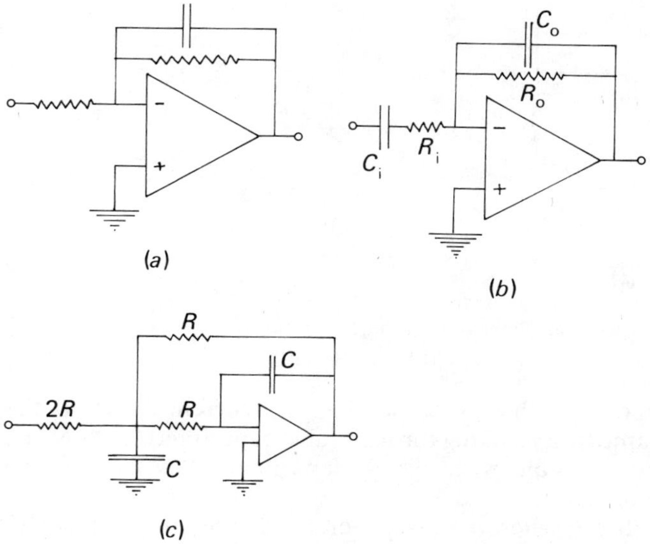


**Figure 6.6** Filter circuits. (*a*) gives an exponential lag, (*b*) can act as a simple bandpass filter and (*c*) will give a second order high cut filter

blocking capacitors to give low frequency cutoff, or a combination of the two, as shown in Figure 6.6(*b*), to give a bandpass filter between frequencies $1/2\pi$ root $(C_i R_i)$ and $1/2\pi$ root $(C_o R_o)$. Such circuits give roll-off rates corresponding to first order systems, i.e. 6 dB/octave. A second order performance with an effective damping of about 0.7 of critical can be obtained with a circuit of the type shown in Figure 6.6(*c*), but more sophisticated circuits are better 'bought-in'. The use of such filters is discussed later; typically 24 dB/octave, i.e. fourth order filters are used but 125 dB/octave 'brick wall' filters can be obtained, albeit expensively. The major problem with high performance cutoff filters occurs in their response to impulsive signals which produce 'ringing'.

## 6.4 Low impedance and charge amplifiers

Some devices, such as photocells, generate a current during operation and this charge must be allowed to dissipate relatively easily when the light is removed or the voltage stays 'high'. It is then essential to have an amplifier that will respond to the minute currents generated, of the order of nanoamps, and yet have an input

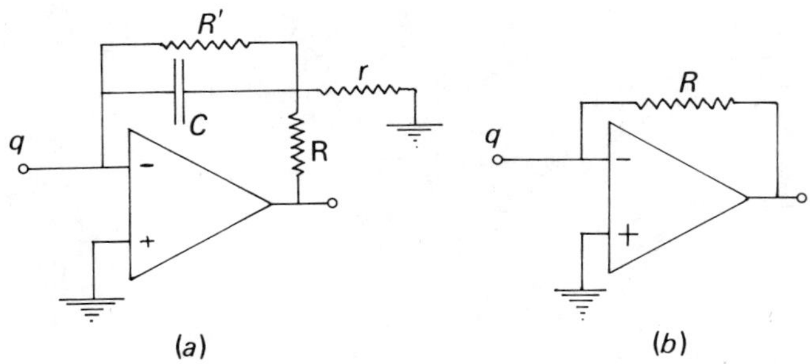


**Figure 6.7** (*a*) Charge amplifier with zero input impedance. (*b*) Circuit for jerk measurement

impedance of less than 1 kΩ so that the current flows to the amplifier rather than leaking through cable capacitance. At 1 MHz, for example, a cable with 200 pF capacitance has an effective impedance of 800 Ω.

Piezoelectric accelerometers generate charge proportional to acceleration and although it is easy to allow the charge to raise the voltage on the self-capacitance of accelerometer and connecting cable, change of cable will alter capacity and hence sensitivity. Also, bending of the cable will alter capacity. The preferred solution is to use a charge amplifier with a zero input impedance obtained by connecting an operational amplifier, as shown in Figure 6.7(*a*).

Any charge developed at input 'sees' earth potential, and hence zero input impedance, as it is immediately neutralised by charge from voltage change across the capacitance $C$. An input charge $q$ will require a voltage change $q/C$ across the capacitor and this voltage change is obtained by a change $q\ (R + r)/Cr$ at amplifier output. The additional parallel resistor $R'$ is solely to give stability against DC drift and is typically 100 MΩ which with 2000 pF will allow operation down to 1 Hz; $r$ must be small compared with the impedance of $C$.

The circuit in Figure 6.7(*b*) has a purely resistive feedback and so again has zero input impedance but an output that is proportional to input current, i.e. rate of change of charge. When used with an accelerometer the output is proportional to rate of change of acceleration, i.e. $d^3x/dt^3$ or 'jerk'. Alternatively, the circuit is useful with photocells.

## 6.5 Displacement, velocity, acceleration or jerk

An accelerometer, the most commonly used transducer for vibration, can measure acceleration, provide 'jerk', as mentioned

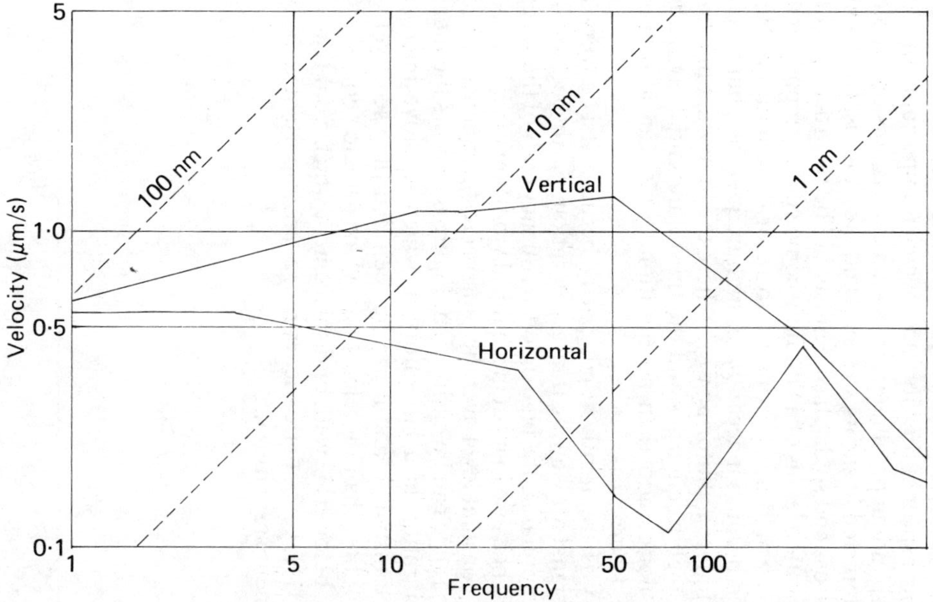


**Figure 6.8** Measured ground vibration. Velocity as a function of frequency

above, or its output can be integrated to give velocity or displacement. There is often confusion as to which parameter is 'best' for recording, and puzzlement as to how the choice should be made.

There are no absolute rules but some straightforward guidelines simplify the choice. The effect of differentiation of a signal with a mixture of frequencies is to emphasise the upper frequencies while integration enhances the lower frequencies. We can then assess the engineering aspect of the vibration, which is important, and choose the corresponding parameter, thus on a car suspension problem, if we are interested in spring stressing and life, it is displacement that governs stress and so displacement should be recorded. In contrast the acceleration levels of a vibrating conveyor are important and in this case it would be appropriate to record acceleration. 'Jerk' is rarely used because it accentuates high frequencies, but it is useful for some failure monitoring work where the important information is at very high frequencies and the third differential gives a better signal to noise ratio than acceleration; 'jerk' can also show up impulsive occurrences.

Velocity is frequently used, partly because it is a very much 'cleaner' signal for viewing than acceleration, partly because it does not over-emphasise the high or low frequency ends of the spectrum, but mainly because in most mechanical vibration work it remains at a roughly constant level over the range of test frequencies, and thus the whole range can be easily seen. Figure 6.8 shows the measured ground vibration in a laboratory in which sensitive microscopes were to be installed, and it can be seen that the use of velocity is more suitable than acceleration which would not show the excitation below 5 Hz well, or displacement which would show little above 100 Hz, yet both frequency ranges are important in the design of equipment suspension.

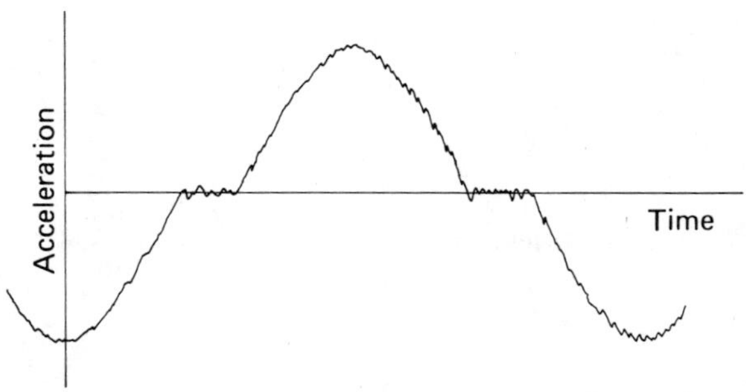


**Figure 6.9** Acceleration trace showing backlash effect whereas double integration to displacement would give sine wave

When force vibration is used to test a structure, the requirement for a machine tool or crystal cutter is that displacement controls, but if a very large structure such as a ship's hull is being tested it has been found experimentally that for much of the time the hull behaves as a mass and that acceleration per unit force, i.e. effective mass, remains roughly constant; it is therefore customary to plot acceleration or 'effective mass' against frequency. Occasionally when investigating a structure a tester may well plot displacement to obtain the main characteristics and resonances of the structure and then change to acceleration if he is looking for the non-linearities that betray slideway movements or insufficiently tight bolted joints. Backlash that is not noticeable on a displacement trace can be very clear on an acceleration trace as illustrated in Figure 6.9.

## 6.6 Bridge systems

Small changes in resistance, inductance or capacitance (frequently less than 0.001%) are common and cannot be measured directly at vibration frequencies; thus a bridge system is used. The classic Wheatstone bridge approach for resistors is shown in Figure 6.10. When used with strain gauges it is usual to have all four arms of the bridge 'working', since there will rarely be less than two 'live' and two compensating gauges. The detector may be either a high impedance voltage device or a low impedance current device. Approximate balance is achieved by separate high value resistors in parallel with the bridge arms. The detector should be able to measure down to microvolts since the voltage drop across 120 Ω with a permissible current of 20 mA is 2.4 V. Thus 0.1% strain gives 4.8 mV change. Measurement to 1 microstrain is required, giving only 4.8 $\mu$V.

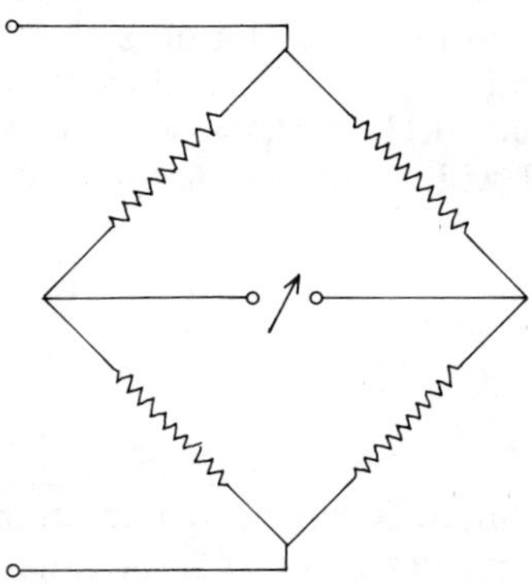

**Figure 6.10** Basic Wheatstone bridge for DC

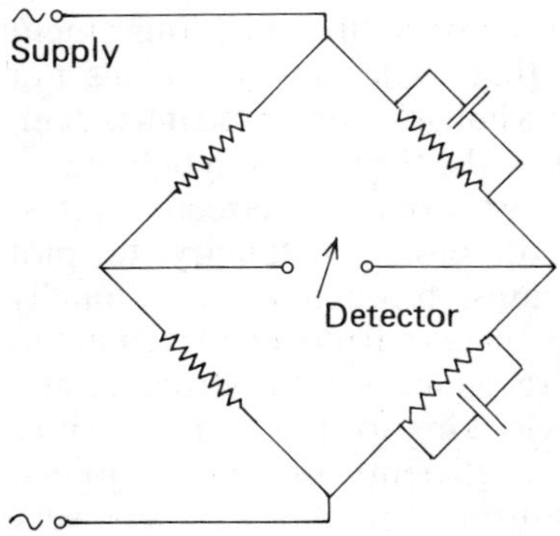


**Figure 6.11** Wheatstone bridge for AC

The same principle applies for AC bridges, as indicated in Figure 6.11, but balancing is more complicated because to achieve a balance both in-phase and quadrature components must be correct. Alternating voltage is applied and so an alternating voltage is produced at the detector regardless of whether an impedance increases or decreases from the zero setting. The difference lies in whether the detector alternating voltage is in phase with the bridge supply or is 180° out of phase with the supply. A detector that is sensitive to phase is required and must be set to respond to the correct phase, i.e. so that it will respond to capacitance variations or to resistance variations according to the transducer being used.

It is obvious that AC systems must be used when inductive or capacitative transducers are used, but although less obvious, there are two valid reasons for their use in resistive strain gauges. Firstly, electronic amplifiers are prone to DC drifts of their zeros and it is difficult to get good long-term stability with a DC system, whereas drift is not at all important for an AC system since it is the carrier frequency that contains the steady state information. Secondly, and more important, most spurious noise in instrumentation systems occurs at mains frequency and harmonics up to 2 kHz, so that if a carrier at 8 kHz is used, filtering at 8 kHz ± 2 kHz will allow information through up to 2 kHz, yet will eliminate mains linked noise completely.

## 6.7 Phase sensitive detectors

The output from an AC bridge circuit or a balanced induction system, such as an LVDT, is a wave at carrier frequency, modulated

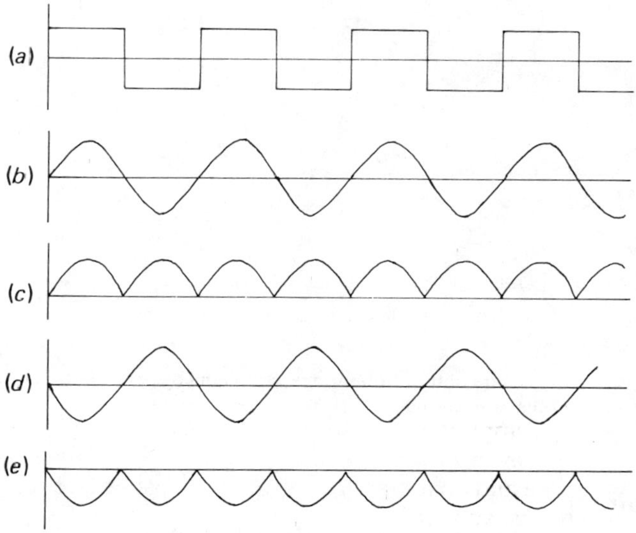


**Figure 6.12** Operation of phase sensitive detector. The square wave (*a*) is in phase with the original carrier wave to the bridge. (*b*) is in phase and so switching gives (*c*), whereas (*d*) is out of phase and gives (*e*)

proportional to movement from a 'balance' position. To decide whether the signal is in phase or out of phase with the original carrier, the detector switches the signal in synchronism with the original carrier; it is simplest thought of as a reversing switch driven by the carrier.

Figure 6.12 shows the effect. When the signals are in phase, both halves of each cycle are positive, and when they are out of phase both halves are negative. Multiplying of the two signals together will give the same effect, though the waveform will be different. The resulting signal is smoothed with a low pass filter which is typically at about one-quarter of the carrier frequency.

Figure 6.13 shows the result of a badly set up system where the output has a large component at 90° phase to the 'reversing switch', and thus the smoothed output signal detects little of the signal variations from the bridge. The 'switch' typically operates at 8 kHz and is a simple transistor circuit. In some equipment there is no subsequent smoothing, the output is only intended to feed a display meter which acts mechanically as a smoothing device; extra filters will be needed if vibration information is required. The filtration associated with smoothing the signal gives the characteristic large phase lags associated with AC systems.

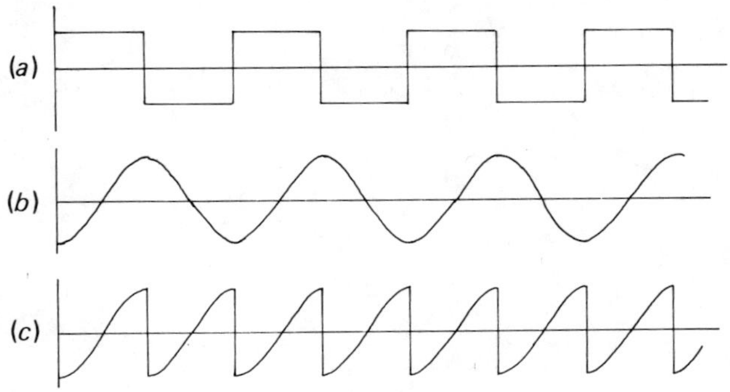


**Figure 6.13** Effect of quadrature output from bridge. Switching by (*a*) on output (*b*) gives (*c*) which smooths to zero value

## 6.8 Enveloping

When the signal of interest from machinery is at high frequencies it may be uneconomic to record or analyse the original vibration signal. A technique that can be used in this case is to filter the signal to remove all unwanted components and then to record only the envelope of the resulting modulated sine wave.

Bandpass filtering presents no problems and with a 25 kHz signal the lower and upper cutoff frequencies might be set at 23 and 27 kHz. Extracting the envelope can then be achieved with the circuit shown in Figure 6.14(*a*) where a diode charges up a capacitor at peak amplitude and slow discharge through a resistor gives (Figure 6.14(*b*)) a relatively large area under the curve and thus a signal that is easier to detect than the original signal, which might

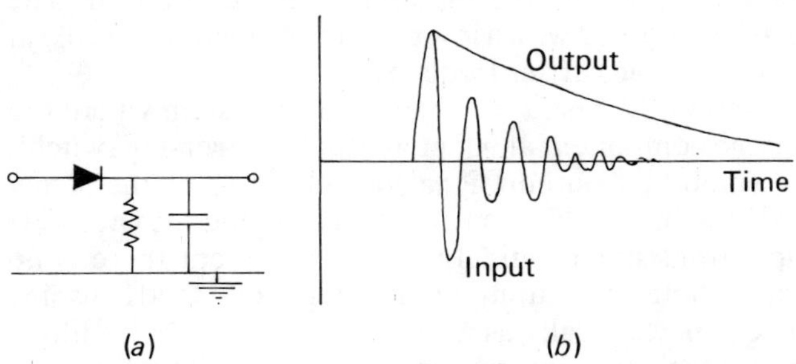


**Figure 6.14** (*a*) Simple rectify and decay enveloping circuit. (*b*) Input and resulting output

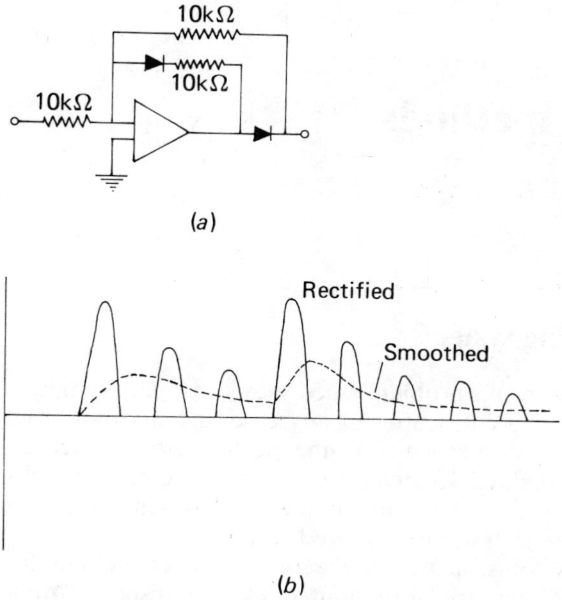


**Figure 6.15** (*a*) Circuit for improved rectification to overcome practical diode limitations. (*b*) Resulting waveform after rectification and subsequent smoothing

last only one-tenth of a millisecond. This method is very useful for general purposes but suffers from a major disadvantage in that it is non-linear because it does not discriminate between a single pulse and several pulses occurring close together. In some of the more sophisticated analyses it is important to retain linearity and therefore the circuit shown in Figure 6.15(*a*) should be used. This gives the rectification required and a low pass filter is then used to smooth the wave, as indicated in Figure 6.15(*b*).

Chapter 7

# Recording methods

## 7.1 Direct writing systems

Steady signals present no problem since they can operate meters or be viewed on an oscilloscope; in any case, signals should be monitored on an oscilloscope for unexpected effects such as transients or overloading. Unsteady signals must be recorded for subsequent study or analysis and in most cases some form of mathematical processing will be required.

Direct writing systems, such as pen recorders, are convenient for giving 'hard copy' but are very limited in two respects: their frequency range is typically 75 Hz at most and paper width is only about 50 mm per track of information; more important is the limitation that the information in the signal cannot easily be converted back into voltage form for analysis so that only visual inspection is possible. This applies equally well to the high speed direct writing devices such as ultra-violet galvometers and ink drop writers. In one a magnetic field rotates a small coil carrying a mirror and in the other an electrostatic field deflects an ink drop, 100 $\mu$m diameter, charged electrostatically. Frequencies can be very high in both cases, well above what is needed for mechanical vibrations.

Photography of the face of an oscilloscope used to be a popular method of recording transients but it is expensive, very time-consuming and again cannot easily provide a voltage variation for subsequent analysis.

Present-day techniques of tape or digital storage have relegated direct writing devices to a role in which their main function is to provide a clear trace for report purposes, for which the hot wire on heat sensitive paper gives a very clear trace and is much used. Another use is for direct readout on slow moving devices such as surface roughness recorders and gear measuring machines.

## 7.2 Tape recording

Until recently tape recorders dominated the field of vibration recording and, despite advances in rival digital systems, are still the single most important method for storing information in large amounts.

A typical standard instrumentation tape recorder will, using approximately 1000 m reels of half-inch tape, have a specification that covers most requirements. Two methods of recording are used and can be mixed in one machine with differing frequency ranges. With a tape speed of 60 inches per second, direct recording can achieve an upper frequency limit of about 500 kHz and a lower limit of about 300 Hz. The same tape speed, using frequency modulation of a carrier to encode the information, will have an upper limit of 40 kHz but will operate down to DC; a reel will last only 10 minutes at top speed but this is sufficient for most purposes. Reducing tape speed reduces the upper frequency limits proportionately but the lower frequency limit on direct recording does not come down below 100 Hz if very high accuracy is required. There is normally a 64 : 1 range in speed down to $^{15}/_{16}$ inch per second.

Fourteen channels of information can be recorded on half-inch tape and a voice track is an asset for easy identification of tests. A typical signal to noise ratio is 48–50 dB, so that a recording at 1 V level would have a 'floor' at about 3 mV; exactly how this is interpreted is imprecise, so it is as well to carry out a check if noise is critical, as the 'noise' is liable to be at specific frequencies.

The great advantages of tape recording are the vast quantities of information that can be stored easily and cheaply, the high data rates possible, and the ability to replay at different speeds. At top speed, 500 kHz on 14 channels would correspond to a digital rate of 14 Mbytes per second on 8-bit resolution or twice this rate on 12-bit resolution. A single reel of tape, costing only a few pounds, can hold the equivalent of $9 \times 10^9$ bytes of information. Replay can be used in a more convenient frequency range by altering tape speed. Normally high frequency information would be slowed down to speeds suitable for computer input, but in a few cases information is obtained at $^{15}/_{16}$ ips over many hours then replayed fast for analysis.

Double recording of a signal is sometimes employed if frequency or amplitude ranges are too large for a recorder to handle. A very wide frequency range of, say, 30 Hz–150 kHz, as encountered in rolling bearing vibration, can be met by recording the same signal on both an FM channel and a direct record channel. Wide ranges of amplitude can occur with signals such as train track vibrations or passing vehicles; here the technique is to record the signal on two channels set to different sensitivities. Part of the time the high sensitivity channel will be saturated but the other channel will have valid information and, conversely, when the signal is low the low sensitivity channel only will have noise on it.

Double recording helps in the particular case of a large low frequency signal such as a once per revolution signal with a very small high frequency signal superposed on it. The signal will be recorded directly, giving the low frequency component but losing the high component in noise, and will also be passed through a high pass filter to eliminate the large low component before further amplification and recording on a second channel which may be directly recorded. Alternatively, the signal can be 'skewed', as mentioned in Section 6.5, and acceleration rather than velocity recorded, so that components over the range of frequencies are at roughly equal amplitudes.

Tape recorders are fairly linear and have a good frequency response. The principal source of error is usually the scaling between input and output signals, e.g. an exact voltage in does not give the correct standard voltage out. For this reason a standard calibrating signal should be recorded on the tape. It is also advisable to record some form of once per revolution marker on one channel to give accurate synchronisation if rotating machinery is involved and if very accurate averaging (see later) is required, then a marker channel with 1024 pulses per revolution is advisable; this can be derived from an encoder or slotted disc.

A slight, hidden, advantage of tape recording information is that the recording process can act as a low pass filter and so help prevent aliasing effects in subsequent digital processing if tape speeds are chosen to suit the cutoff required.

## 7.3 Transient recording

Digital recording of information can be used for capturing transients or can be the input into analysis computations; it can follow tape recording after speed change into a convenient frequency range.

Transient recorders usually have limited capacity, typically 4000 data points and normally work at 8-bit resolution, i.e. recording on an integral scale from 1 to 256 (or 0–255). (This is usually quoted as 0.4% resolution and is adequate for viewing on an oscilloscope or for general analysis purposes, but it is not good enough for accurate work). They are very good, as the name implies, at capturing transients such as impulses and giving a semi-permanent storage from which:

1. The data can be read out quickly and repetitively to give flicker-free viewing on a standard oscilloscope, turning it into a 'memory scope' for visual check or analysis.
2. The data from 'good' results can be read out in analogue form at a speed which is sufficiently slow to drive a pen-recorder for 'hard copy' or into a subsequent analyser.
3. The data in digital form can be transmitted directly to a computer or analyser for processing.

Transient recorders may be separate items but more commonly are built in to give a 'memory scope' and, as prices have dropped, are increasingly used for initial setting up and monitoring purposes as they can 'catch' signals that are at frequencies too high for the eye to follow on a normal oscilloscope.

Some caution is needed due to aliasing problems, discussed later. The input channels of the units do not have high quality antialiasing filters built in and thus severe aliasing problems can be encountered, particularly when there are large high frequency oscillations on the signal from an electronic instability or pickup. This can of course be checked by the operator by going well up the frequency range and determining whether or not large signals are present. The reason for the lack of filtration is partly due to the very occasional requirement for deliberate detection of beat frequencies between sample frequency and a high frequency oscillation, but is mainly due to the very high cost of good filters which would be more expensive than oscilloscope and memory put together.

Transient recorders can work very quickly with typical sample rates of over 1 MHz and in theory have sufficient point storage for analysis purposes. In practice if analysis is required they are not normally used. Either a dedicated frequency analyser is used or the information is fed to a high speed buffer which may store up to 0.5 M points at frequencies up to 1 MHz. Some preliminary processing may be carried out at high speed in the buffer in order to reduce the amount of information before it is fed to the (relatively slow) computer.

## 7.4 Digital recording

Digital recording now dominates vibration work simply because all information must be converted to digital form if any of the mathematical analysis techniques are to be used.

A signal is scaled and filtered, then sampled in typically a microsecond and is 'held' while a high speed analogue-to-digital converter generates the digital version. As above, 8-bit resolution is somewhat coarse for most work and thus 12-bit (i.e. integers 1–4096) is commonly used. Negative voltages are not used and so, for a symmetrical signal, zero input voltage would correspond to about 2048 output. On a standard $\pm 1$ V peak signal this gives 0.5 mV resolution, usually adequate. The penalty for the extra resolution is that the information takes up more computer space; as systems such as the IEEE 488 data transfer currently work on an 8-bit (1 byte) basic unit, 12-bit resolution takes up 2 bytes and there are 4 unused data bits in each 'point'.

Once digitised, the points are fed into memory in sequence; the choice is then between immediate processing of the information or simply storage for subsequent processing. High speed powerful computers or special purpose equipment will operate 'in real time', i.e. will immediately carry out the calculations and present the information. In contrast, a small computer such as a personal computer is relatively slow and will not be able to store and calculate quickly enough; it stores the data points and then calculates relatively slowly, taking perhaps several minutes. If a computer language such as Basic is being used, operation is very slow.

Until recently, memory was a considerable limitation on digital methods. Random access in small computers was limited to 32 or 64 k points so that, after leaving room for operating systems and computation routines there might only be space for some 10 k 2-byte data points. Four channels of recording up to 2 kHz frequency, each requiring 5 kHz sampling rates would allow only 0.5 seconds operation before the memory was full. This would be satisfactory for steadily running machinery but not for irregular processes. Special purpose vibration analysers can now store 10 k points and analyse any block of 1024 points, but memories on small computers have moved up into the >1 M region at low cost. A problem arises in that transfer to floppy disc memory is relatively slow, so that although 10 seconds of operation generating 200 k points can be stored immediately, the transfer to disc might take several minutes.

## 7.5 Pre-processing

Tape can store vast quantities of information, but digital methods give better resolution and so it is worth while investigating whether the amount of information in the signal can be reduced by some form of initial processing, whether digital or analogue, to ease digital storage problems and computation time.

The most powerful reduction process is time-averaging (see later), which can be used for rotating machinery where processes occur regularly at once per cycle. A typical example is a four-stroke engine where a cycle corresponds to two revolutions. Provided the phenomenon of interest is the regular or average behaviour the information from, say, 128 cycles can be compressed into one cycle. This method works very well in such systems as gearboxes where the regular vibration associated with one shaft speed can be separated from the other shaft speeds by averaging. The resultant number of points, probably 1024, give high accuracy and resolution but do not require either the memory capacity or processing time associated with the original data. What is required is an averager that can operate quickly with the necessary fine discrimination on output.

Another technique is enveloping, where it is known that the only vibration of interest occurs at, say, 20 kHz in the form of short bursts of vibration. The signal is filtered so that only the 20 kHz remains and is then enveloped, usually by some form of rectification and either a low pass filter or a decay circuit. The resulting signal may be a 'blip' occurring at a repetitive frequency of about 100 Hz and with significant harmonics up to 1 kHz so that the amount of information to be stored per revolution has dropped by a factor of 20.

Frequently, information in a signal is reduced using Fourier analysis: 1024 data points may be processed into 400 lines of information on amplitude and 400 phase items; this is of little advantage, but if only a few of the 400 frequencies are significant the remainder can be ignored and there is a large reduction in the amount of data but the original signal can still be regenerated.

## 7.6 Method selection

There are no clear cut rules on the choice between digital and tape methods of recording but some tendencies can be mentioned.

1. If there is a high data rate, e.g. 14 channels of 20 kHz information, then tape is likely to be better.
2. If the data rate is too high, above 1 MHz sample rate of a single channel, then tape is not fast enough.
3. Large total quantities of information, of the order of many megabytes, will need tape storage.
4. Minicomputers can only process information slowly and are thus unsuitable for high frequency or high volume work.

5. A tape recorder is much more portable and will withstand harsher environments than a normal computer and memory system.
6. Digital resolution at 12 bits corresponds to nearly a 20 dB better signal to noise ratio than tape, but 8-bit resolution is less efficient. However, if averaging is employed, 8-bit sampling can give the equivalent of 12-bit resolution (see Chapter 12).

## 7.7 Aliasing and sampling

Digital storage of information can hardly be discussed without acknowledgement of the problems of 'aliasing' or 'picket-fence' effects.

The problem at its simplest is shown in Figure 7.1. A pure sine wave at frequency $f$, which is above the sampling frequency $s$, will show points that lie on a sine wave of frequency $f - s$. Frequencies just above the sampling frequency will thus be detected as low frequencies and, similarly, frequencies just below the sampling frequency will appear as low frequencies at frequency $s - f$.

In general terms any frequency $f$ sampled at frequency $s$ will appear as frequency $\pm f \pm ns$ where $n$ is any integer. Subsequent analysis should ignore any frequencies above about 0.4 s but if any wave gives a difference frequency between $\pm 0.4$ s it will appear in the sampled signal. There is conceptually no difference between a positive and a negative frequency because they both give a vibration of the form $a \cos \omega t$ since $\cos \theta = \cos(-\theta)$.

The term 'folding frequency' is sometimes used in this context because the observed spectrum after sampling can be obtained by folding the original spectrum at frequencies $s/2$, $s$, $3s/2$, etc. This is because $f$ and $s - f$ appear as the same frequency and it is as if the region from $s/2$ to $s$ is mirrored between $s/2$ and 0. Similarly, $s - f$ and $s + f$ give identical samples and a mirror image about frequency $s$. Figure 7.2($a$) shows the spectral lines obtained from a series of sine waves of decreasing amplitude with frequency, and

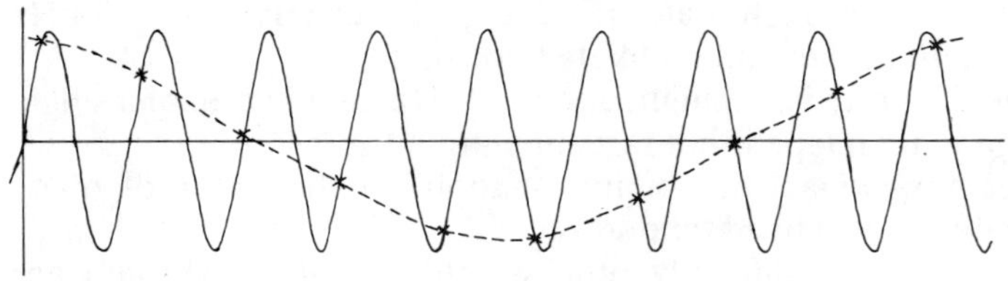

**Figure 7.1** Effect of sampling a high frequency signal to give an 'alias' frequency

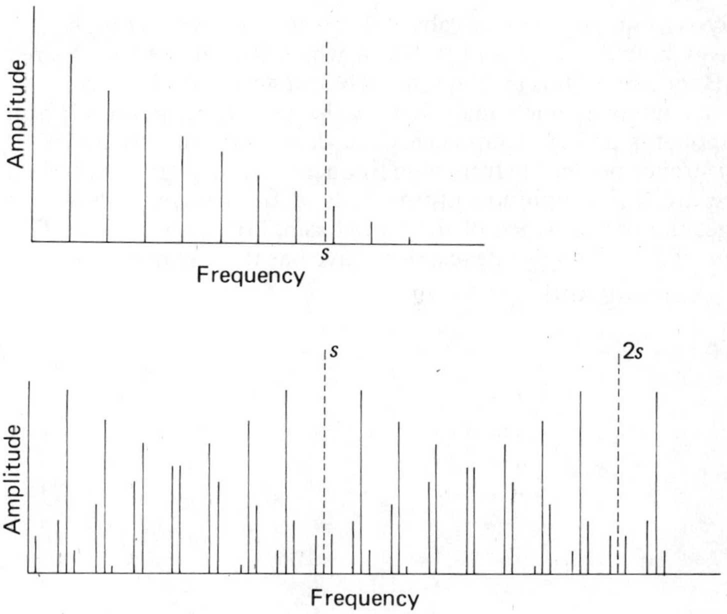


**Figure 7.2** (*a*) Actual frequency analysis of a signal. (*b*) Measured possible analysis of the same signal after sampling at frequency *s*

Figure 7.2(*b*) shows the corresponding observed spectral lines after sampling at frequency *s*. A silly answer may be obtained exactly at frequency *s*/2 since samples are consistently at the same two points on the sine wave and may even all read zero.

Sampling theory [1] gives the information that a sampling rate *s* will give information on a sine wave up to frequency *s*/2, but this performance cannot be achieved in practice because of the impossibility of differentiating between waves just above and just below *s*/2. The problem can only be solved by filtering out the components above frequency *s*/2 so that they cannot appear in the sampled signal. If we wished to analyse up to 0.4 s we would require a filter that passed waves at 0.4 s without attenuation, but that eliminated all frequencies above 0.6 s.

Elimination is a rather absolute term but in practice we would wish to reduce the unwanted components to less than 1% of the components of interest, and as both might start at the same amplitude we need a 40 dB reduction for little more than half an octave. Filters having an 80 dB per octave performance can be obtained but since they are expensive it is easier to reduce the analysis frequency range to about 0.25 s so that there is roughly 1.5 octaves between 0.25 and 0.75 s in which to achieve the desired 40 dB reduction; this is then relatively straightforward.

Fortunately in much analysis work the amplitudes of the spectral components at high frequencies are decreasing rapidly and thus interference or aliasing from high frequencies is negligible but when there are high amplitude components at frequencies outside the range, the performance of the 'antialiasing' filters is critical. This occurs if a high amplitude 'carrier' wave has to be eliminated.

Chapter 8

# Vibration excitation

## 8.1 Introduction

Thus far we have ignored the source of the vibration of interest. If machinery is running steadily with a regular vibration output, it is a straightforward exercise to use the machine itself as the source of vibration, although this is not practicable in many cases.

Problems arise because either the vibration of interest is destructive, is inherently unstable, or is irregular such that it is difficult to carry out detailed analysis because operating conditions are varying continuously. In a machine tool the excitation may vary in amplitude and direction while the natural frequencies move with changes in the stiffnesses and effective masses of overhanging slideways. It is then usually better to determine the frequencies of interest rapidly and to investigate mode shapes and characteristics at leisure, using a controlled excitation whose amplitude and phase are known. Each method of excitation tends to have a clearly defined set of uses.

## 8.2 Methods available

1. *Electromagnetic moving coil (electrodynamic) vibrators*: these are more powerful versions of a moving coil loudspeaker. A cylindrical coil of low resistance, usually a single layer of winding, moves in the direction of its axis with the coil interacting with a steady radial field, as shown in Figure 8.1. The magnet, whether permanent or DC excited, is relatively heavy and cumbersome but the coil, wound on an aluminium or laminated plastic former, is light. Its weight will be typically about 1% of the thrust developed so that accelerations of the order of 100 *g* can be achieved on the coil. The coil suspension may be a flexible plastic spider or a complex rolling strip system in large vibrators. A rough rule of thumb is that an electromagnetic vibrator will weigh about the same as the peak force it can attain. Frequency range is from DC to 10 kHz for small

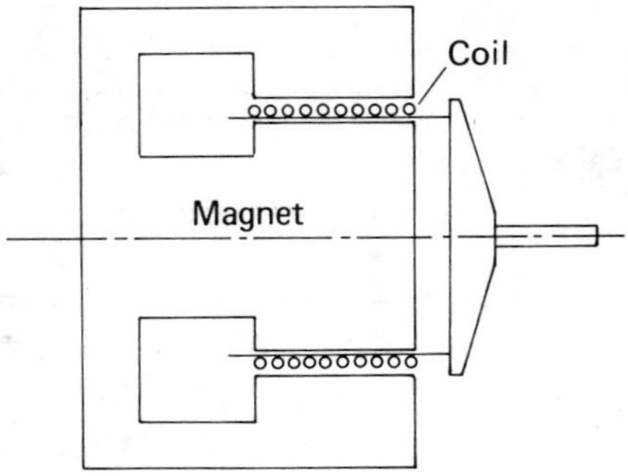


**Figure 8.1** Section of moving coil vibrator with cylindrical coil in radial magnetic gap

(10 N) vibrators and rather less for large vibrators. Very large vibrators can be obtained with forces of the order of tonnes but costs are high, of the order of £10 per N thrust.

2. *Hydraulic vibrators*: these use a small electrohydraulic servo valve of the type used in aircraft controls to direct high pressure oil (at 200 bar) to either side of a ram which is connected to the test object. A force of 1 Tonne can be achieved with a ram of only 25 mm diameter. The limit of performance is dictated either by the response speed of the valve which will behave as a damped single degree of freedom system with a natural frequency of about 200 Hz or by the limiting flow rate through the valve. A valve will have a flow rate of the order of 20 l/min (5 g/m) and so with a flow of ⅓ l/s and a ram area of 500 $mm^2$, the maximum velocity will be 0.6 m/s for a 1 Tonne ram. At a frequency of 200 Hz, this corresponds to an amplitude of 0.3 mm, so at high frequencies only very stiff structures such as machine tools can be tested but at low frequencies, e.g. car suspension frequencies, high amplitudes pose no problem. The valve and ram are small, typically have a mass of less than 2 kg, and so can be fitted easily into a small space although a large, expensive and usually noisy hydraulic power pack is needed. As oil is relatively compressible, care must be taken to keep ram internal volumes very low.

3. *Mechanical out-of-balance systems*: a motor with an eccentric weight will give a rotating force and if two such weights are geared together the result will be an oscillating force of $2mr\omega^2$ where $mr$ is each out of balance. Forces can be high but the amplitude is proportional to the square of the speed, which is not convenient, and frequency is limited by motor speed of typically 5000 rpm, i.e. 80 Hz.

4. *Piezoelectric devices*: these are the converse of accelerometer crystals and generate a change in size when charge is applied. The strains generated are very small, of the order of $10^{-5}$, and thus a disc of material will give very small amplitudes. An alternative is to use a bar of material polarised such that the bar bends when excited. This gives somewhat larger movements if fixed as a cantilever, but it is of most use if it is attached to a turbine or compressor blade which will be excited in bending. This form of excitation is sufficiently robust to withstand high acceleration fields and gives minimum disturbance to air flow.

5. *Magnetostrictive devices*: magnetisation of some nickel alloys gives a significant length change with a strain of the order of $10^{-5}$. The corresponding small movement is amplified by exciting a standing wave in a tapered probe at ultrasonic frequencies of about 20 kHz. The vibration is normally used to generate cavitation in a liquid to give small shock waves for cleaning purposes, or it can be used with abrasive to cut through brittle materials. Alternatively, sound waves can be generated for such purposes as crack detection.

6. *Electromagnetic pull*: variation of the current through the magnetising coil of an electromagnet will give a variation in force across a gap. The operating force is limited to a maximum of about 0.5 $N/mm^2$ (75 psi), comparable to the pressure exerted by a factory airline. The system is non-linear since force is proportional to the square of magnetic field strength and is a pull force regardless of current direction. The advantage is that the steel which completes the magnetic circuit can move sideways so that is is possible to excite a rotating steel shaft without contact. Some form of Hall probe is needed to measure field strength and hence deduce force and the 'target' must be ferromagnetic; eddy currents can limit the high frequency performance.

7. *Air excitation*: this specialised excitation is for light or non-magnetic objects and uses a Rayleigh loudspeaker principle. A high pressure air supply (at several atmospheres) is controlled by a comb valve; a valve might have gaps of 0.8 mm alternating with slats of 0.8 mm so that a $\pm 0.4$-mm vibration will give the range from fullflow to non-flow and the 'comb' which must be moved, usually by a moving coil vibrator, need not have a mass of more than about 0.2 kg so the frequency range is reasonable. Unfortunately the noise level is very high.

8. *Impact testing*: with the advent of fast digital processing, impact testing is becoming much more popular. A hammer applies a rather arbitrary shape of pulse which will contain all frequencies to a

greater or lesser extent. The applied load is either measured directly with a piezoelectric load cell or is deduced from the hammer head deceleration, measured by an accelerometer. A softer, heavier, hammer, sometimes with a plastic insert, will emphasise the lower frequencies. Testing is rapid but the power input level at any particular frequency is low relative to the peak force attained.

## 8.3 Relative advantages

The electromagnetic moving coil vibrator dominates the field of testing because it has such a wide frequency range and easy control. The other methods of steady vibration will usually only be considered if a moving coil is not suitable. As a rough rule, hydraulic excitation should be assessed if forces exceed 1000 N or if forces above 100 N are required within a small space or with negligible additional mass; frequency range is limited and so it is unusual to find hydraulic vibrators working above 200 Hz, even though it is possible.

The very low mass of cantilever piezoelectric 'benders' allows them to be used in rotating blades where they will not distort airflow and can withstand centrifugal fields.

Rotating shafts are easiest excited by having a bearing on the shaft with its outer surface excited by a moving coil vibrator, but if this is not allowable the electromagnetic pull or air blast type must be used.

The magnet of the moving coil vibrator is heavy and its mass can rule out its use if light plastic models of a structure need to be given a relative excitation between two points. The simplest solution in this case is to use two vibrators, supported separately from the model, to act in antiphase to give a relative force.

Impact testing is a method that requires relatively complex signal capture since there is no steady state, but it is very convenient for rapid testing. If, for example, only a short time between trains is available for testing railway track response, the lack of necessary attachment points and the short testing time in the danger area are great advantages. Impact testing also has the advantage that it is easy to move the application point to ensure that all modes are excited and this may be of great use when reciprocal testing; it can, however, only be used approximately square to a surface.

A problem encountered with both impact and moving coil excitation is due to unwanted restraints or excitations. When a hammer impacts there is not only the main impact force in the 'correct' direction but also a friction force at right-angles, typically up to 0.25 of the main impact. The friction force is unknown and is variable in size and direction, so a series of impacts will give differing results if there is any significant interaction between impact and perpendicular directions. This effect often appears as a puzzling 'noise' in the results and is usually averaged out (hopefully) by taking several impacts. If the effect is consistent it is less likely to be noticed by the experimenter (or computer) and so is unlikely to be detected.

A corresponding problem can occur with moving coil excitation if the excited structure is attempting to twist in the region of the attachment but is prevented from twisting by the rigidity of the coupling to the vibrator. The coupling should be necked down or pivoted to prevent this occurring in those cases where twisting is likely.

Hydraulic exciters are compact and rigid and thus may impose both twisting restraint and sideways movement restraint unless care is taken.

## 8.4 Mass compensation

Accurate work at high frequencies requires an exact knowledge of the force applied to the system. The force may be measured by a strain gauged link but there will still be a small additional mass between the point in the link at which the force is being measured and the point at which it is applied to the structure. When the inertia forces associated with the additional mass are significant a correction must be made. This can be done either by measuring the acceleration and calculating the amplitude (and phase) of the force, or by running the vibrator without connection to the test specimen and recording the inertia forces that are developed. The signal may be corrected by subtracting a voltage proportional to acceleration using an operational amplifier and adjusting the scaling factor so that no net force is recorded when the specimen is disconnected.

## 8.5 Machine excitation

Machines that have out-of-balance, meshing gears, plunger-type pumps, etc., have an internal source of excitation that can be of considerable use. The noise source in a large gearbox is the relative motion between the teeth and it is impossible to get vibrators at the interface, so that only possible source of excitation is the gear meshing operation itself.

Fixed speed operation, as when a synchronous AC motor is used, does not give the necessary information. If there is a variable speed drive the change in speed gives a change in excitation frequency analogous to sweeping a vibrator through its range. This allows the determination of the shape of the resonance response of the system using overlapping harmonics of the excitation. Taking a particular example: for a drive motor which can vary in speed from 1200 to 3000 rpm and which drives a gearbox with 20 teeth on the input gear, tooth frequency will range from 400 Hz up to 1 kHz. Overlapping this range, second harmonic will give 800 Hz–2 kHz, etc. Provided the load torque is held constant the excitation at tooth frequency and harmonics should remain at constant amplitude and thus a plot of the output vibration level will be a plot of the response of the system. In general, the amplitude of excitation will not be known and the relative size of first and second harmonics, though not known, can be deduced from the region between 800 Hz and 1 kHz where the ratio of outputs from the two harmonics is the ratio of their input amplitudes, since the response is a function of frequency only.

This approach depends on the excitation remaining at constant amplitude regardless of speed, an assumption that is true of a gearbox under constant torque or a plunger pump at constant pressure, but that is not true if load varies. Typical curves obtained are shown in Figure 8.2.

Alternatively, the output vibration from such a system can be measured at a particular frequency such as a resonance of the system as the speed varies. Variation of output is not due to the system since response is fixed at a given frequency and so is due

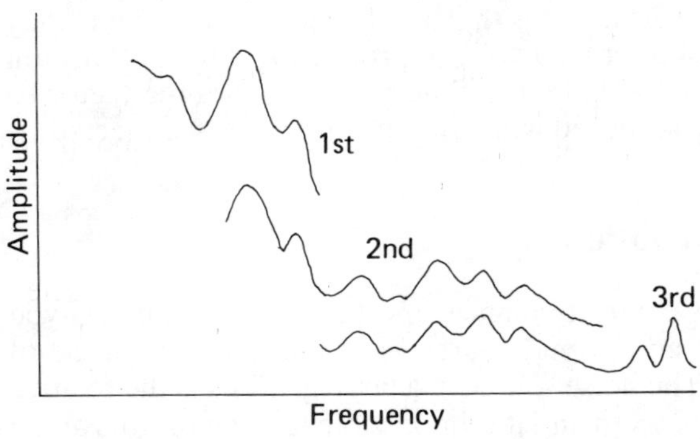


**Figure 8.2** Typical experimental curves obtained of vibrations at first, second and third harmonics of pump plunger frequency

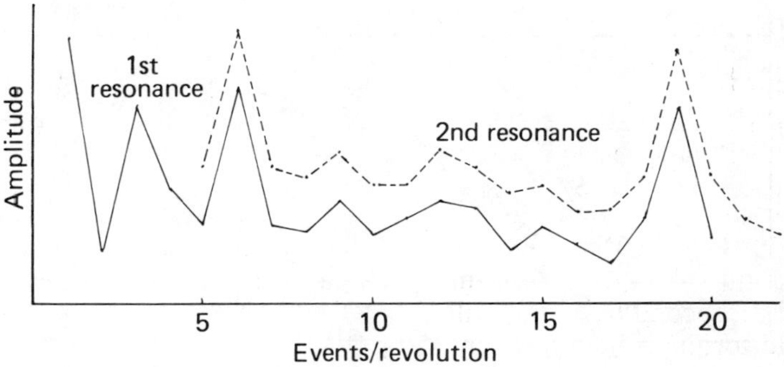


**Figure 8.3** Determination of relative sizes of excitation components at multiples of 1/rev by measurement of vibration at two particular resonance frequencies

solely to vibration of input to the system. The simplest method of plotting is to specify output as a function of 'events per revolution' to give the type of plot shown in Figure 8.3 and this allows comparison of the relative sizes of the input components. In this example there is strong forcing at 1/tooth, i.e. 19 'events per revolution' and sidebands, and some forcing at frequencies such as 6 per revolution due to pitch errors on the gear.

## 8.6 Waterfall displays

These are a particular form of machine excitation and give a rapid overall impression of the response of the system and its excitation frequencies. The computer takes a series of frequency plots as machine speed is decreased and displays them simultaneously on the screen with a vertical positioning proportional to machine speed. This is equivalent to drawing a Campbell diagram, as shown in Figure 8.4; the lines at fixed frequency are the resonant frequencies and lines radiating from the origin correspond to excitation frequencies, e.g. at 19 × per revolution, etc. Peaks of output will occur whenever an excitation line meets a resonance.

The corresponding 'waterfall' display is shown in idealised form in Figure 8.5 and shows similar effects with excitation frequencies steadily decreasing with speed and high peaks at coincidences. Unfortunately 'waterfalls' and Campbell diagrams tend to be plotted the opposite way round, with frequencies horizontal in the case of the waterfall and vertical for the Campbell diagram.

Limitations for waterfall displays are the same as those for conventional cross-plotting as described above. In practice, waterfall

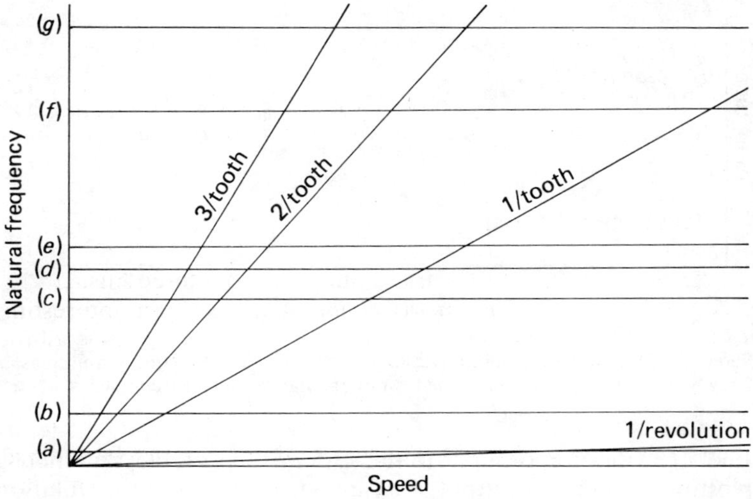


**Figure 8.4** Campbell diagram to see where excitation frequencies meet natural frequencies

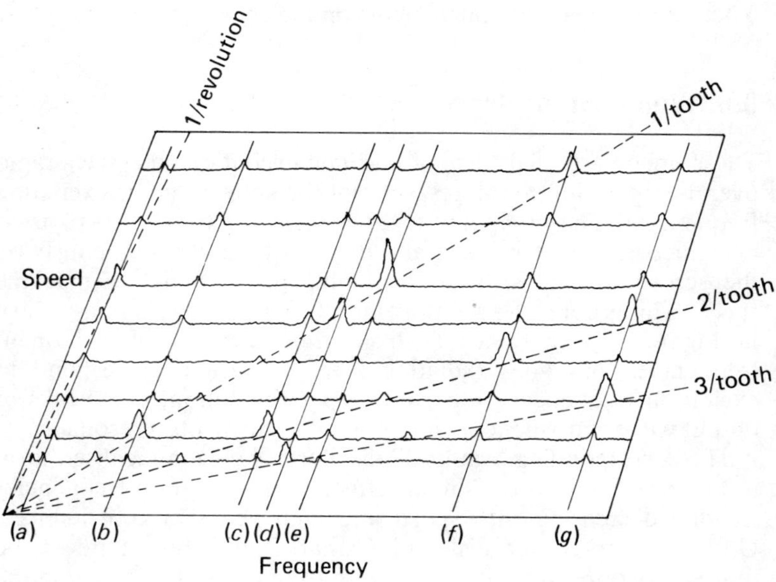


**Figure 8.5** Waterfall display showing a series of frequency analyses as speed is reduced. Frequencies (*a*)–(*g*) correspond to natural frequencies of the system. 1/t is tooth frequency, 2/t second harmonic, etc.

displays are often obtained on rundown when the load torque has been removed from a system; thus the levels of excitation may have changed greatly from the operating torque condition, and non-linearities may appear. Despite this limitation, the displays are very useful for a qualitative picture of vibration patterns.

## 8.7 Reciprocal theorem

The reciprocal theorem is occasionally encountered, usually in structures analysis, and is often considered as an interesting 'oddity'. It is, however, an extremely useful and powerful tool for practical testing and can save a great deal of time and effort on site; it may often also save computational effort.

In its simplest, static, form the reciprocal theorem states that in any *linear* system if a force applied at position 1 gives a deflection at position 2, the same force applied at position 2 will give an equal deflection at position 1. As both force and displacement are vectors there must be due regard for direction. In Figure 8.6, if force **F** applied at 1 gives displacement **e** at 2 then **F** at 2 gives **e** at 1. The same is true dynamically, both in amplitude and phase so that, though it sometimes appears unlikely, it does not matter whether we vibrate at 1 and measure at 2 or vice-versa.

Typically we can measure vibration with an accelerometer of mass 20 g in a volume of less than 1 ml, but excitation may require a vibrator mass of 20 kg and a volume of 1 l. Also, it is very much easier to move an accelerometer than to move a vibrator. The reciprocal theorem allows us to minimise the practical difficulties of installing vibrators in difficult areas by reversing the measurement.

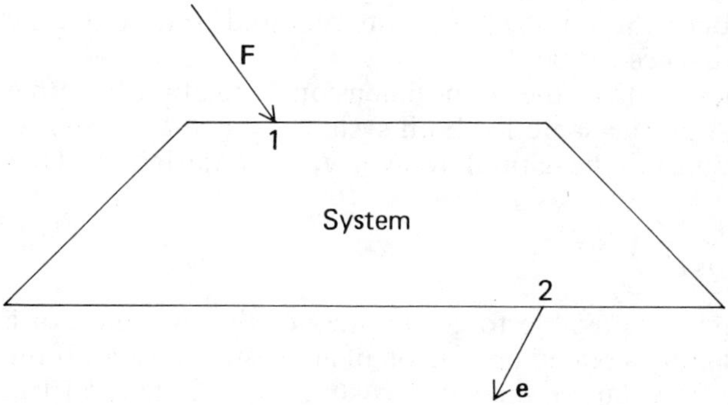


**Figure 8.6** Reciprocal theorem. The deflection **e** obtained at 2 due to **F** at 1 would be the same at 1 if **F** were applied at 2

The source of a vibration in an installation may be an inaccessible bearing housing on an engine or gearbox and the vibration of interest may be at a mounting foot which is clear of obstructions. Reversing the test allows easy coupling of the vibrator at the mounting foot while the accelerometer is fitted at the bearing and the reciprocal theorem is invoked.

A similar approach may be used when a mode shape is being determined by impact testing on a structure. In this case the impacting hammer is moved more easily than an attached accelerometer. Instead of impacting at the vibration source and attaching accelerometers round the structure it is simpler to attach one accelerometer at the 'source' and move round with the impact hammer.

## 8.8 Model testing

Model testing has rather gone out of fashion as finite element techniques have been promoted enthusiastically and it is certainly correct that once a system has been set up it is much easier and quicker to check the effects of changes on a computer model than on a plastic scale model. Both methods fall down in that they are unreliable for joints and slideways but give sensible results for the type of structure, usually large, where joints are not important.

Plastic small scale models may, however, still have a use when the system is so complicated that computing cannot cope with the volume of information. This is likely to be the case in a large structure that is subdivided into many individual cells, each of which is subjected to local bending waves as well as direct compression and shear. A plastic model, though not easily modified, is probably the best method once the number of points required to describe the structure approaches 10 000.

Scaling rules follow the non-dimensional relationship that $\omega^2 L^2 \varrho/E$ must be the same for both systems; $\varrho$ is the density, $E$ (or $G$) the modulus, $\omega$ the natural frequency, and $L$ the length. Thus

$$\frac{\omega_1}{\omega_2} = \frac{L_2}{L_1}\left(\frac{\varrho_2 E_1}{\varrho_2 E_2}\right)^{1/2}.$$

Occasionally it is possible to get an idea of the dynamics of a design if it is simply a scaled up version of an existing design. If the new design is 25% larger than the existing one, its natural frequencies will be 20% lower. Running an old gearbox 25% faster than normal should then give a good idea of the vibration response to be expected on a larger design since the ratio of tooth frequency to natural frequency will be the same in both cases.

## 8.9 Sweep, random, chirp and impulse

The question is often asked as to which test method is the 'best' method of determining the response of a system. As so often happens there is no simple answer. The basic approaches are:

1. *Slow sweep*: sinusoidal input at constant amplitude with a sufficiently slow sweep to ensure that the transients die away and the response is steady state. This involves slow frequency change near very low damped resonances. A very rough rule is that it takes a system $Q$ cycles of vibration near a resonance to 'settle down' to a steady state where $Q$ is the dynamic amplification factor of the resonance.
2. *Random input*: white noise, or an approximation to it, is used to excite the system so that all frequencies within the operating band receive equal input power, and the ratio of output to input power for each frequency band gives the required transfer function.
3. *Impulse*: an impulse, if short, is the same as white noise in that it supplies equal amplitude at all frequencies such that a ratio of output to input power in each frequency band gives the transfer function. A longer impulse, using a softer hammer, tends to reduce the high frequency content but this merely alters the input frequency distribution; the computed transfer function will still be correct.

Method 1 is relatively slow and may take several minutes so that if very many runs are required the operator's patience may be strained. It is the method used with an analogue approach and is less suited to a 'real time' digital approach. A random input has all frequencies simultaneously and thus gives a very fast answer digitally; the impulse method is similar.

The major difference between 1, the sweep, and 2, the random input methods lies in the power levels. Vibrators are limited in output force and so when random noise is used there is little power in a given frequency band. The full theory is cumbersome [1] but a simplistic approach suggests that if a 400-line spectrum is used, with input power spread evenly into 400 bands, a given band can only contain $\frac{1}{400}$ of the limiting power output of the vibrator. In comparison, a slow sweep gives full power at each frequency. The corresponding 20 to 1 reduction in amplitude (actually approximately 30 to 1) can give signal to noise ratio problems and may prevent the use of the random approach in noisy environments. The situation can be improved by, say, averaging 128 spectra but this loses the original speed advantage of the random approach. If there is a suspicion that the system is non-linear at small amplitudes, then the white noise input should not be used.

An alternative 'random' approach uses a signal that is 'pseudorandom'. The signal is a square wave so that the vibrator is being used to high power all the time, unlike white noise, but the spacing of the square wave is controlled to give roughly equal amplitudes of vibration at the frequencies to which the analyser is set.

The impulse method, 3, nominally overcomes the low power limitations of random noise by greatly increasing force levels. Peak forces with impacts are typically 10 000 N rather than 100 N with electromagnetic vibrators. However, although force levels are much higher, the impulse is for a short period and so total input energy does not rise correspondingly. The $\times 100$ increase in force levels is countered by the time of application being less than 1% of the time during which the output must be analysed to catch all the decay curves, thus signal to noise ratios may still be troublesome.

Chirping was devised to overcome the signal to noise problems of white noise by compromising between sweep and random approaches. (Note that, typically, 'noise' is being used in two completely different senses, both as desired random input vibration and unwanted spurious output vibration.) A sinusoidal input signal is used and is swept through the frequency range very rapidly, giving a single 'chirp'. At a resonance there is insufficient time for the amplitude to reach the steady state condition and so visual inspection of the output will only show the beginning of a build up of amplitude. However, the system response will still be computed correctly; in practice repeated sweeps are often used. Chirp testing, by removing the necessity for waiting until amplitudes stabilise, is very much faster than the conventional slow swept sine approach.

Whichever method is used, the fundamental laws of testing cannot be avoided; accurate repeatable results require either that input energy levels in a given frequency band are appreciably greater than noise levels, or that long testing times must be used to average out random noise effects. Much is made of the superior speed of 'real time analysis' with a random input when selling equipment, but as this test is only part of a vibration investigation, time may not be important unless many tests are being made.

One aspect of the choice between exciter and impact methods depends on possible cross-coupling effects, discussed in Section 8.3.

If the system is likely to respond strongly to sideways friction forces, impacting hammers should not be used, but if it is important that no torsional restraint be imposed, a normal vibrator attachment should not be used.

Chapter 9

# Analysis methods and frequency analysis

## 9.1 Introduction

Analysis of vibration signals can be carried out by a very wide range of methods. The novice tends to assume that there is a single 'correct' method of processing signals and this belief is sometimes reinforced by enthusiastic salesmanship. At one end of the scale there are problems where a glance at the vibration trace on an oscilloscope is all that is needed, while some monitoring techniques require a very complicated filtering, time averaging, frequency analysis, line elimination, resynthesis and then statistical distribution (Kurtosis) analysis.

In some cases such as in machine tool structure rigidity work the vibration signal is of little interest and the dominant requirement is the measurement of mode shape for structure improvement.

The important factor is the type of information required and this must be decided in detail before analysis commences. As a very simple example, for repetitive systems such as engines or pumps it is sometimes the average over many cycles that is important, while at other times it is the variation from cycle to cycle that matters; the approach to the two problems is very different. Similarly, the steady vibration from a gearbox is important for detection and location purposes but it is the unsteady part of the vibration that gives a clue to tooth damage.

Some rough guidelines can be stated but it should be emphasised that each problem should be approached with care, as it is all too easy to destroy information unwittingly. Frequency analysis is of most use when there is a small *steady* component of interest hidden in a noisy signal. Time averaging is of use in any complicated machinery with shafts running at different speeds since it can extract the components associated only with that shaft. Cepstrum, the frequency analysis of a frequency analysis, is useful where there is modulation of a vibration due to such causes as fluctuating loads. Kurtosis, the second moment of area of the statistical distribution of signal amplitude, is useful for detecting some types of damage in the presence of random noise. Correlation, which gives information on the time delay of one signal relative to another, is useful for checking transmission paths of vibrations through large structures.

## 9.2 The raw vibration

It is surprising how often it is possible to see a large and complicated set up for checking vibration, with no provision for viewing the original 'raw' vibration trace. In theory, viewing on an oscilloscope should not be required but in practice a look at a vibration trace can often give valuable clues when something unexpected is happening.

Human beings are very good at extracting the essential character of a vibration from the trace and will rapidly and automatically absorb whether the vibration is steady or unsteady, wrongly set on amplitude, high frequency or low frequency and whether a 'steady' signal is distorted. Such information can be obtained by, say, a twin channel frequency analyser but it will take longer and will be less specific in 'noticing' errors.

A second stage of detection is easily accomplished by eye in determining whether an unsteady signal is merely irregular or is coming in short bursts associated with impacts, and whether such bursts are coming regularly or intermittently. This type of information is invaluable for selecting the best method of numerical analysis. Distortion of a signal in the form of harmonic content may not give a clear idea of trouble, but if the waveform is as shown in Figure 9.1, there is very likely to be play or backlash in the set up in the central position in the first case, or saturation in the second case.

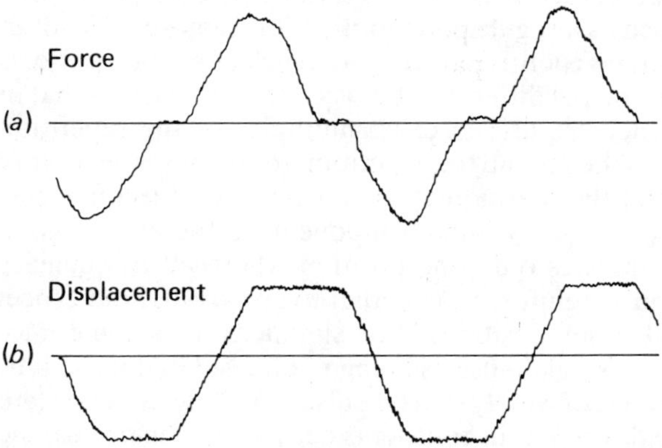


**Figure 9.1** Effect of phase of harmonic content on observed waveform

A variant on oscilloscope viewing is the use of a pen recorder to capture the signal, but as frequency response is limited to about 50 Hz, the signal may first need to be slowed down with a tape or digital recorder. Ultra-violet recorders were much used in the past but they are cumbersome so it is more usual to employ a memory oscilloscope and print out onto a pen recorder.

If vibration is arriving irregularly, then an immediate question arises as to whether it is occurring repetitively at a particular point in an operating cycle. A once per revolution pulse from a transducer is invaluable and triggering an oscilloscope from such a pulse, with a sweep time of about one revolution, will quickly identify regular occurrences. Use of a memory oscilloscope is probably not advisable for this due to the limited number of samples in the sweep and the correspondingly limited frequency response when on 1/revolution speed.

## 9.3 Frequency analysis

The most commonly used method of analysis is separation of the signal into its components at a range of frequencies or frequency bands. Analysis may be called frequency, Fourier, Fast Fourier, band, wave, power, spectral, power density, spectral density, harmonic, real time, third octave, etc. All frequency analysis eventually depends on Fourier's original approach, which starts from the basic assumption that any waveform, however complicated, whether repetitive or not, even containing discontinuities, can be described or generated as the sum of a series of sine and cosine waves at different frequencies.

This fundamental equality can be seen to be true for any repetitive waveform such as a regular saw-tooth but it is more difficult to conceive when a non-repetitive occurrence, such as a single pulse, is of interest. The principal difference between the two cases is that in the first the frequencies are all exact mutliples of the repetition frequency, while in the second the repetition frequency is very, very low (i.e. zero) and the 'harmonics' then occur at all frequencies. Figure 9.2 attempts to show some components in the two cases; in the second case there is only one position where all frequencies agree in phase and so reinforce each other to give a pulse but cancel each other at all other positions. For simplicity it is sometimes easiest to think of a single pulse as being not an isolated pulse but, rather, one of a series of widely spaced pulses. As long as the system has time to settle down, i.e. its transients decay between pulses, the response will be the same for one pulse only as for one of a string.

(a)

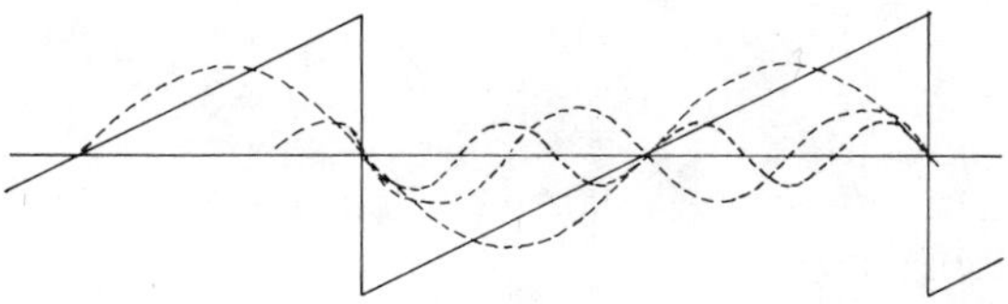

(b)

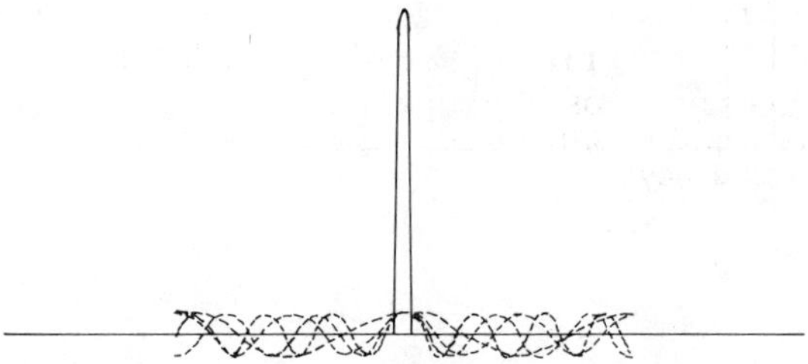

**Figure 9.2** (*a*) A few of the Fourier components of a regular saw tooth wave, and (*b*) of a single pulse

## 9.4 The two types of spectrum, conversions

Much confusion arises in frequency analysis due to the inherent differences between the two types of spectrum which occur. The two types of spectrum require a different presentation of the end result and since equipment can only be operated to present the information in one or the other form, the user may become muddled.

Rotating machinery such as plunger pumps, gearboxes, petrol or diesel engines produces a vibration that is at very clearly defined, exact frequencies. A simple example is an out of balance force which occurs precisely once per revolution and at no other frequencies; hydraulic noise from a reciprocating pump will occur once per plunger and harmonics but not at intermediate frequencies. The type of distribution of power of these vibrations is shown in Figure 9.3(*a*) where all power is at specific frequencies; they are usually referred to as *line* spectra or discrete frequency sources.

In contrast the waves hitting the legs of an oil drilling rig not only have variable amplitude but also have variable frequency and occur at all frequencies. Noise associated with turbulence is similar in that it approximates to 'white' noise and occurs at all possible frequencies within the range of interest. The corresponding plot of the distribution of power is shown in Figure 9.3(*b*) and is called a continuous or distributed power spectrum.

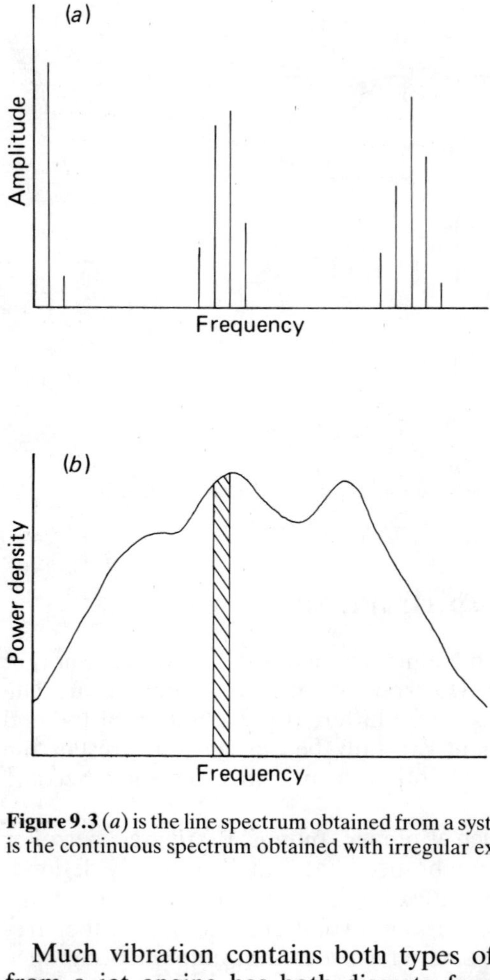


**Figure 9.3** (*a*) is the line spectrum obtained from a system with exact frequencies. (*b*) is the continuous spectrum obtained with irregular excitation such as wave impacts

Much vibration contains both types of distribution; the noise from a jet engine has both discrete frequencies associated with blade passing frequencies from the compressor, and a continuous spectrum from the jet turbulence. When landing there is a characteristic single frequency whine, whereas full power gives a distributed 'roar'.

A radio signal will have a continuous spectrum within a very small (±10 kHz) frequency range centred on a carrier frequency of the order of 100 MHz. An overall frequency analysis would simply show a single line at 100 MHz, but it is the continuous distribution just around the carrier which contains the information of interest, whether speech, pop or classical music.

## 9.5 Bandwidth and power

The idea that the vibration power in a system can be presented as a plot against frequency follows directly from the basic Fourier idea that any vibration signal can be broken down into the sum of steady vibrations. We can take the original signal, separate it into its components and then plot the amplitude of each component against its frequency.

This approach works well with a line spectrum; a square wave can be analysed as a series of odd harmonics whose amplitude decreases proportional to frequency and gives the plot of distribution as shown in Figure 9.4. An original signal of $\pm 1$ V, for example, gives a fifth harmonic of amplitude $4/5\pi$ V and it can be checked, relatively easily, that the total power in the series sums to the power ($1\,V^2$) in the original signal. There is no uncertainty about the frequency involved since if the basic frequency is 60 Hz the fifth harmonic will be at 300 Hz.

However, a continuous spectrum presents difficulties since, mathematically speaking, *all* frequencies are present, though possibly within a limited frequency band. If, then, a frequency has a finite amplitude, it has finite power; and an infinite number of frequencies, each with finite power, give infinite power. Since this is an impossibility, each frequency must have only zero power and we cannot plot the power of a single frequency. This forces us, mathematically and from common-sense practicalities, into measuring the power in a frequency band rather than at a particular frequency.

Conceptually we pass the original signal through a 'perfect' filter which passes the band of interest without attenuation and

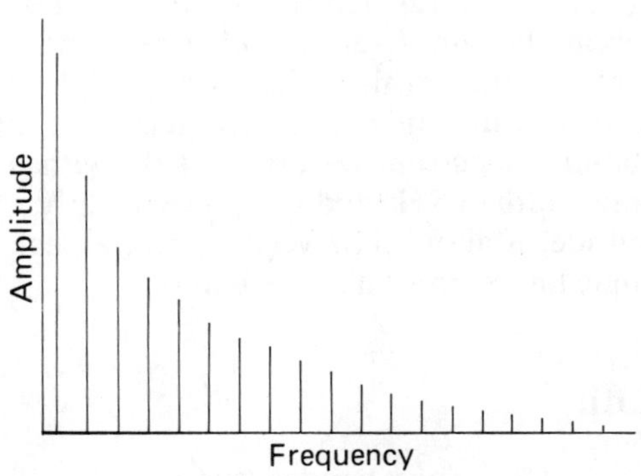


**Figure 9.4** Frequency analysis of a square wave

completely eliminates all other frequencies, then we measure the power, usually root-mean-square, left in the band. However narrow the band, the infinite number of frequencies within it gives a finite power as expected. Doubling the width of the band doubles (roughly) the power measured and so it is logical to specify the power per unit bandwidth as the measured parameter. Provided the spectrum is continuous, the power per unit bandwidth will not vary greatly if we alter the bandwidth. The resulting figure is called the power spectral density (PSD) and the units, assuming volts are being measured, will be quoted correctly as $V^2/Hz$. This is the parameter that is calculated by digital techniques and is the figure of interest for random vibration or continuous distribution of power. The value of the bandwidth used is not critical unless there are rapid variations of power with frequency.

Occasionally a 'frequency distribution' will be quoted in units of $V/(Hz)^{1/2}$, obtained by taking the square root of the PSD figure. There seems little advantage in using this variant unless standard convenient bandwidths, such as 1 or 100 Hz, are involved; comparison with individual spectral lines is simplified for cases where the two are mixed.

Line spectra do not fit happily into a PSD approach since there is a fixed finite amount of power at a particular frequency. Reducing the analysis bandwidth does not reduce the power observed and halving the bandwidth doubles the PSD. For this type of spectrum the requirement is the measurement of amplitude, i.e. volts, since this is the factor that remains constant regardless of bandwidth. It is also the factor that is likely to have meaning for the 'customer' since volts will be a direct measure of the acceleration or displacement that is causing trouble and so may be related easily to requirements. An out-of-balance vibration of 5 $\mu$m has a physical concept that is more easily understood than a spectral density of $10^{-12}$ $m^2/Hz$, though they may produce about the same overall level of vibration.

Conversion from one presentation to the other is achieved by multiplying the PSD by the bandwidth and taking the square root to give the amplitude. A 400-line spectrum covering 2 kHz with a reading of 3 $V^2/Hz$ has a bandwidth of 5 Hz and so a power of 15 $V^2$, corresponding to an amplitude of about 3.87 V (rms) and a peak amplitude of 5.48 V if a single line is present in the band.

## 9.6 Choice of bandwidth

A question often asked is: 'What is the best bandwidth for analysis?' and, as usual, there is no 'best' answer. Many of the detailed problems are similar to those which occur in filtering signals and which are discussed in Chapter 13. The more general question again depends on the particular engineering questions that are being asked about the process.

A traditional approach, using fixed analogue filters, was to specify octave bands, and this is still a good method for an initial look at a spectrum covering a very wide frequency range. The vibration from many engineering processes approximates to 'pink noise', the term applied to a continuous spectrum, which, unlike 'white noise' has constant power density per octave instead of per Hz. This type of vibration, typified by the vibration from a ball bearing, can spread over four decades from 20 Hz to 200 kHz, and displaying the whole span of the vibration on a modern digital constant bandwidth device is no great help since nearly all the power is concentrated in the first line which is 500 Hz wide.

The general approach is to have a look over the whole range of vibration using a coarse bandwidth and then to narrow the bandwidth at 'interesting' parts of the range. There may, however, be other information such as noise information which allows immediate focussing onto a frequency that is known to cause trouble. Typically, there will be a single frequency such as pump plunger frequency which is of interest but if, as usually happens, there is variation of speed or amplitude the signal modulation (see Chapter 10) will give sidebands. Some problems are only concerned with the overall power at approximately plunger frequency and require a coarse bandwidth to give a single value for the overall power. Alternatively, the modulation may be of interest and it is necessary to have fine resolution to separate the sidebands out; if a 90-tooth gear is under investigation with once per revolution effects at 60 rpm, the sidebands are only 1 Hz away from the 90 Hz tooth frequency and thus a bandwidth of less than 0.5 Hz is desirable. A 1% variation in speed (jitter) would swamp sidebands under these conditions and controlled sampling (see Chapter 11) may therefore be needed.

## 9.7 Methods of obtaining spectra

The traditional analogue approach to obtaining a spectrum was to filter, using a resonant circuit with a dynamic amplification factor of up to 50. The frequency of the resonance was varied through the range slowly enough to ensure that 'steady state' was reached and the filtered signal was rectified so that its amplitude could be recorded as a function of frequency. The method was slow since with a high $Q$, waiting for stability of output meant that a single sweep would take about 2 minutes for vibration signals. The bandwidth was poor since 2% represented a practical limit and it was often not possible to separate sidebands.

Digital methods are described by a variety of names, but usually work by the same 'FFT' method. Filtering by digital means is relatively slow and expensive and thus is avoided. The classical Fourier approach would involve taking the sampled signal (typically 1024 points), generating the corresponding 1024 points of a sine wave at a test frequency, multiplying together and summing the results. This operation, repeated for each of the 400 frequencies of the spectrum, would require large memory and would be very slow since computers are relatively slow at multiplication. Addition is a fast process and a method was thus developed to economise on computer power using addition instead of multiplication.

The method is known as the Fast Fourier Transform or FFT; although there is much computer interest in the details of the method, it essentially gives the same answers as the classical Fourier multiplication approach and there is no need for the casual user to understand the difference. The algorithm used works by repeatedly splitting up the sampled signal into shorter sequences and is described in standard books on random vibrations [1, 2]. A possible aid to understanding how multiplication can be turned into addition is to consider what would happen if the classical Fourier approach of multiplying the time signal by a sine wave were replaced by multiplying the original signal by a square wave of amplitude $\pm 1$. This would only involve addition or subtraction of the points in the original signal. The answer would be incorrect since there would be contributions from third, fifth, etc., harmonics in the square wave. If, however, these were known from higher frequency 'multiplications' subtraction would give the required coefficient.

Analysis is sometimes termed 'real time'; this merely specifies that the operation is sufficiently fast that there is not a noticeable delay between the end of sampling and the display of the spectrum on the screeen. The practical alternatives are the analogue method mentioned above or the use of mini-computers which may take an appreciable time to carry out the computations, thus delaying viewing by several minutes. This latter alternative, though much slower, is significantly cheaper than the use of specialist machines but it is normally restricted to a maximum sampling rate of the order of 20 kHz by interface transfer rate limitations.

## 9.8 Windows and limiting bandwidth

So far, no mention has been made of the length of the sample necessary for analysis and it has been assumed that a 'long' trace was available. The reality is that sampling occurs only for a finite number of samples, usually 1024, and that with this restricted number of points there may be only two complete waves present of the lowest frequencies in the whole sample length.

The mathematics of the Fourier process is the same regardless of whether the signal is a transient or is steady and, if presented with a wave as shown in Figure 9.5, the system will analyse that shape. This will give not only an incorrect amplitude but a spurious DC level and the full range of higher harmonics from the sudden changes at the ends.

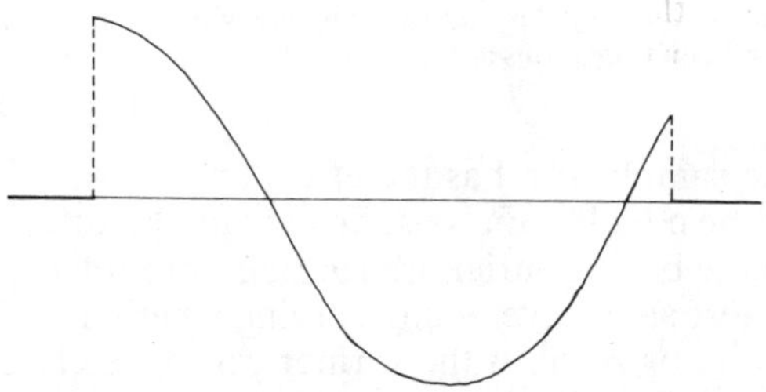

**Figure 9.5** Effect of finite sample length on a low frequency wave

The way of reducing spurious harmonics from the end effects is to ensure that the signal decreases smoothly to zero at both start and finish of the sample. This is achieved by multiplying the original sample by a 'window' which attenuates at the ends to eliminate the discontinuities, as indicated in Figure 9.6. The most commonly used window is a 'Hanning' but several shapes can be used. An irregular signal or a signal where the total sample time is not an exact number of cycles of a repetitive waveform requires the use of a 'window' and this is thus the standard condition for a frequency analyser. If the waveform is a transient, there is no need to smooth the ends since the waveform will start at zero before the impulse and the sample length should be chosen so that the wave has decayed to zero; then for a transient a rectangular 'window' (i.e. no modification of the signal) can be used, but if there is any suspicion that the trace will not be zero at start or finish, it is safer to use a smooth 'window'. The various different 'windows' and their advantages and disadvantages are described in Ref. [2].

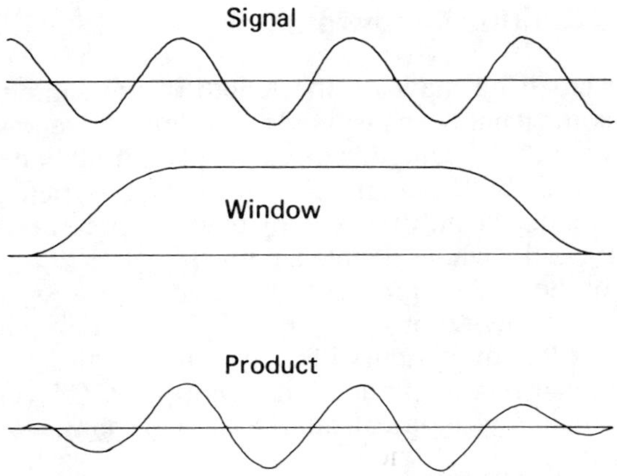


**Figure 9.6** Window effects. The original signal is multiplied by the window shape to give the product which avoids sudden transitions at the ends

The finite length of the sample also has the effect of imposing a limit on the bandwidth of the possible analysis. To explain this effect it is simplest to return to the basic Fourier approach of multiplying the vibration signal by a test sine wave and observing whether the averaged amplitude is finite or zero; in the former case there is a signal present at that frequency; in the latter there is not. Figure 9.7 shows two waves of slightly differing frequency, one a vibration, one a 'test' wave; over the sample time shown, it is necessary for the vibration to become 180° out of phase before the average multiplied value will reduce to zero. A frequency change which only gave one-

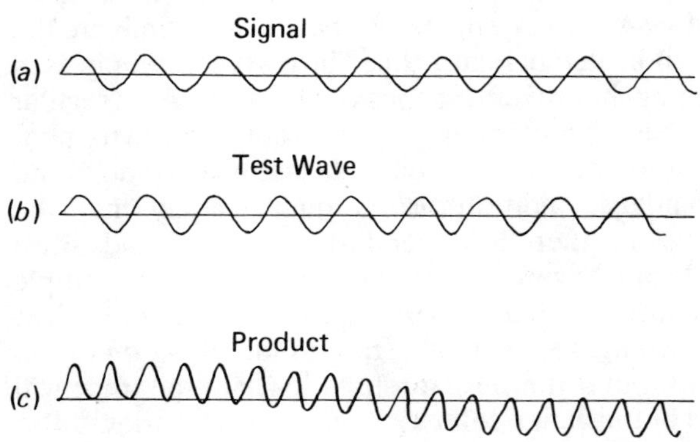


**Figure 9.7** Effect of sample length on detectable frequency differences

quarter of a cycle change within the sample length would give little reduction in observed frequency component. Detailed theory of the 'sample length' effect gives the practical working rule that it is only possible to resolve frequencies if they are a cycle apart in the sample time, so that if the sample time is $B$ seconds the frequency resolution is $1/B$ Hz. This rule is affected slightly by the type of 'window' used with the sample, but for most purposes it gives a very good idea of the limiting resolution.

## 9.9 Repetitive waveforms

Repetitive waveforms contain more information than the corresponding irregular waveforms. Although in both cases typically 1024 points are sampled, with an irregular waveform we do not have any more information than is conveyed by the points, whereas with an exactly repetitive waveform we know precisely what all previous and all future points are for that vibration. Because we have an infinitely long repetitive sample length we can say that only line spectra can exist and these only at multiples of the basic once per cycle frequency.

A problem arises if the repetition frequency does not correspond to the sampling time since a batch of 1024 points might contain 1.6 cycles and the analysis routines could not deal immediately with this. Working by hand (and eye) we would identify the exact length of the cycle and then carry out an analysis solely on the 640 points corresponding to a cycle. These 640 points, repeated to infinity, would allow exact frequency analysis up to, in theory, 320 times per cycle; only integral mutliples of cycle frequency are allowed, so it is not possible to have a power content at a frequency of 2.5 times per cycle. This approach could not be carried out reliably with the normal frequency analyser and thus it is necessary to synchronise the sampling to the machine cycle.

The solution to this synchronisation problem is controlled sampling, which also overcomes 'jitter' problems, and this is discussed under averaging techniques in Chapter 11.

## 9.10 Incorrect amplitudes at borderlines

A problem can arise with standard equipment if a line is present in the spectrum at the cross-over point between two bands in the frequency range (see Figure 9.8). Although in reality there may be a single line with a given power level, the analysis may give two

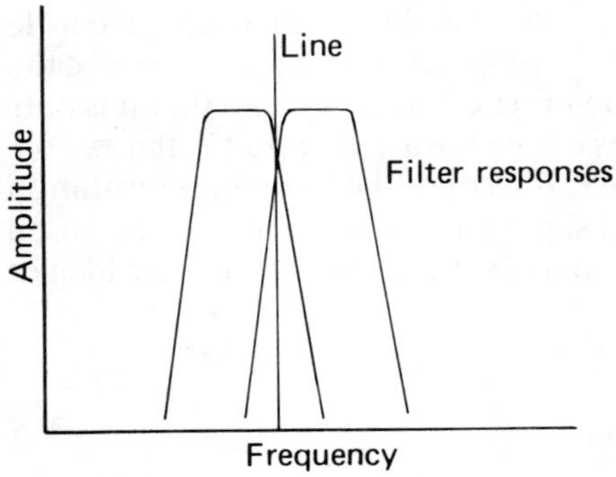


**Figure 9.8** Coincidence of a spectral line with the borderline between two frequency bands

neighbouring lines with the power divided between them and so give a misleading impression of amplitude of vibration. The effect is the same with both analogue and digital filters.

At a particular frequency, say 100 Hz, changing ranges will not avoid the problem since such frequencies will remain on a borderline whatever the range. Addition of the powers recorded from the two neighbouring bands will give an indication of the power of the borderline component but some inaccuracy may be involved.

Avoiding this problem involves a frequency shift in the bands or a fundamentally different approach.

## 9.11 The importance of phase information

Frequency analysis tends to be dominated by the inspection of amplitudes and usually little notice is taken of phase; on some early equipment there is no provision for display of phase.

The most dramatic demonstration of how important phase can be is shown in Figure 9.9. The waveforms in (*a*) and (*b*) can both give equal amplitudes at all frequencies of interest; both a single impulse and random 'white noise' can produce the same distribution as shown in (*c*). The only difference between the two traces is that of phase. In a single narrow pulse the phases of all the frequency components coincide at the pulse (as in Figure 9.2), whereas in 'white noise' the phases are random. As is obvious from the time traces, the engineering deductions are liable to be very different for the two cases since one suggests damage or abnormality while the other is common when there is no damage.

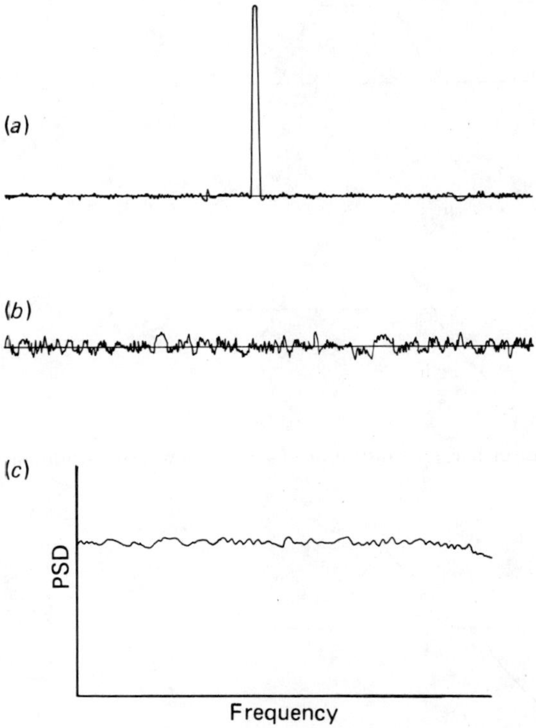


**Figure 9.9** Phase effects. The pulse in (*a*) and the white noise (*b*) can give identical frequency analysis as in (*c*)

Even with a very limited number of harmonics in a wave, the effects of phase can be significant. In Figure 9.10(*a*) and (*b*) there is approximately the same amount of third harmonic present to distort a fundamental, but one trace would indicate saturation or a softening spring instead of the 'bump stops' suggested by the other.

Another variant is shown in Figure 9.11 where the trace shown is from a force transducer and suggests the presence of free play or backlash; similar harmonic content occurring elsewhere in the wave would not lead to the same deduction.

Sophisticated techniques of signal investigation, discussed later in Chapter 13, require preservation of phase information so that after removal of unwanted signals a 'clean' signal can be regenerated, and it is then very important that phase information is not distorted or lost.

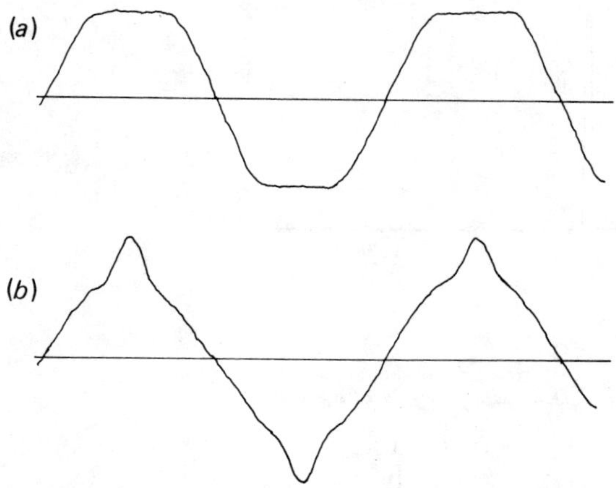


**Figure 9.10** Effects of phase on harmonic distortion of a signal showing very different waveforms

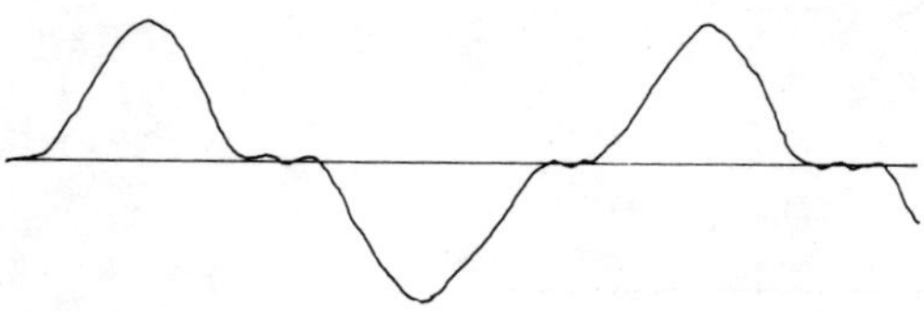

**Figure 9.11** Harmonic distortion due to free play

A further demonstration of the importance of phase (and the attendant dangers of Fourier analysis) occurs in the case of a diesel where there are sharp, isolated impacts between the teeth of the timing gears. The impacts are very sharp, spaced at 1/tooth intervals or occasionally 2/tooth intervals, and usually only occur at one point in the cycle where torque is a minimum.

The results from conventional frequency analysis are completely unhelpful, as they appear as white noise. Display of phase is equally unhelpful whereas viewing the original time signal, preferably triggered from a 1/cycle marker, gives a very clear set of engineering deductions.

Chapter 10

# Modulation

## 10.1 Mechanisms of carrier modulation

The terminology of carriers, modulation, etc., has come from the electrical field and so seems slightly out of place in mechanical vibrations, but some mechanical systems give a nearly regular excitation which is affected (or modulated) by problems in the system.

A good example occurs with gearbox vibration and so can serve as a model for other types of vibration such as plunger pump vibration. In a gear mesh the dominant vibration frequency is usually the tooth meshing frequency or harmonics; an accelerometer placed on a gearbox reducing speed from 1800 rpm to 360 rpm on a 106- to 21-tooth ratio would expect to show a major component at 630 Hz.

Inspection of the vibration trace may show a signal such as that illustrated in Figure 10.1 where the basic once per tooth frequency component varies in amplitude but the frequency is constant. This is typically at once per revolution of the gear and is often associated with manufacturing inaccuracies.

This, the simplest type of modulation, is appropriately called amplitude modulation. Visually it can be seen to be of the form $(A + B \sin mt) \sin ct$, where $c$ is the basic carrier (or tooth frequency) and $m$ is the modulation frequency.

The alternative form of modulation, termed frequency modulation, is illustrated in Figure 10.2. The amplitude remains constant but the frequency varies cyclically so that the waves arrive 'early' or 'late' compared with a steady vibration. This type of vibration is of the form $A \sin(c + \varepsilon \sin mt)\, t$, where, again, $c$ is the basic carrier frequency and $m$ is the modulating frequency.

**Figure 10.1** Amplitude modulation

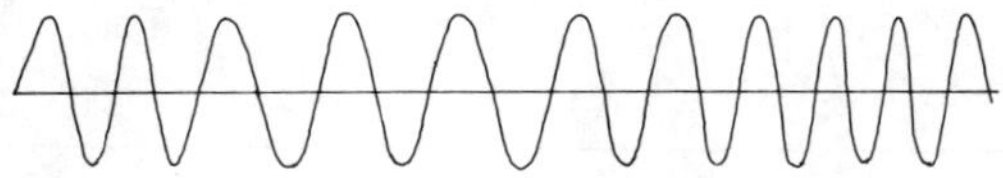

**Figure 10.2** Frequency modulation

Frequency analysis, whatever the method used, has no method of describing varying amplitude or frequency and can only present information as 'steady' sine waves. Mathematically, the first case splits down into three steady components: $A \sin ct$, $B/2 \cos (c-m)t$, and $B/2 \cos (c+m)t$, to give a frequency analysis as shown in Figure 10.3($a$). Very strong modulation gives a relatively small value of $A$ and correspondingly there is no fundamental left, merely the two sidebands. This can appear very puzzling when the frequency of interest does not exist in the Fourier analysis results. The time signal

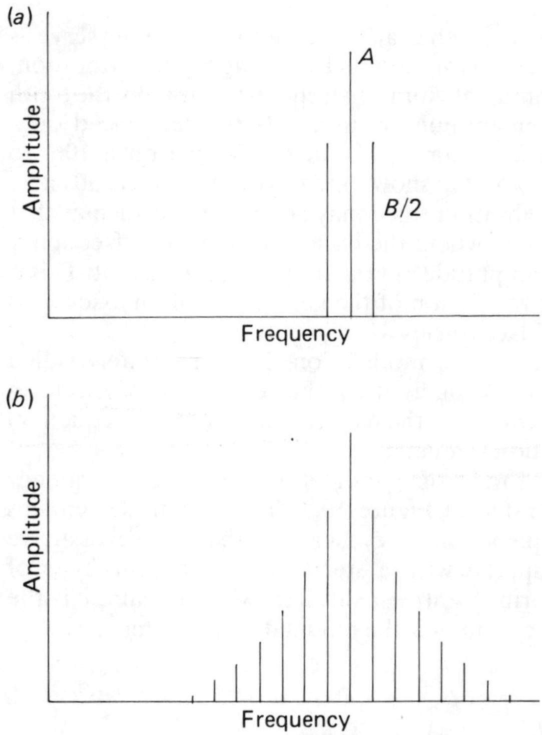


**Figure 10.3** Fourier analysis of amplitude modulation ($a$) and of frequency modulation ($b$)

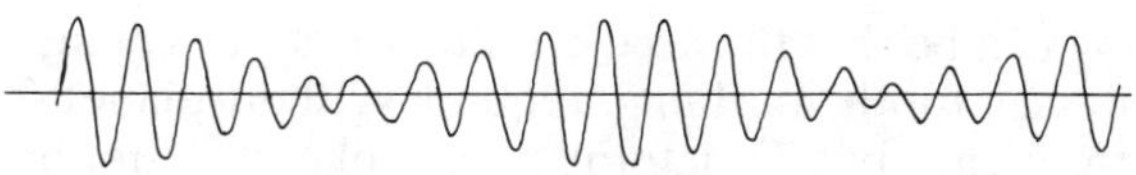

**Figure 10.4** Heavy modulation of signal giving twice per revolution signal amplitude variation

looks as shown in Figure 10.4; a helical gear that has been mounted with its axis misaligned (with swash) will give this type of signal with *two* regions during a revolution with small vibration level.

Mathematically, frequency modulation is much more difficult and instead of a single pair of sidebands it gives the result shown in Figure 10.3(*b*) with an infinite series of sidebands at *m* frequency spacing. Frequently in practice only the first sidebands are distinguishable and the others disappear into the background noise.

## 10.2 Relevance for testing

It is a straightforward procedure to deduce what frequency analysis will result from a given modulation, but the reverse is not true. The presence of background noise in the spectrum of Figure 10.3(*b*) is likely to submerge all but the first sidebands so that there will be no obvious difference between the two types of modulation. This effectively prevents deductions from the amplitude distribution to determine which type of modulation is occurring.

However, if both types of modulation occur simultaneously the result is not a simple addition of power in the first sidebands. For one sideband the effects of amplitude and frequency modulation can occur in phase and thus add, whereas the effects can be 180° out of phase for the other sideband and thus subtract. This double effect can occur in gear systems and produce extreme cases in which one sideband disappears completely. There is thus the possibility that of the three frequencies, all three, any two or just the 'carrier', i.e. tooth frequency, can occur. Alternatively, a modulation can be present that is 90° out of phase and has little effect on sidebands.

In special cases such as in an epicyclic gearbox an unusual effect can occur due to the effects of the movement of the planets round the annulus. An accelerometer fixed on the annulus will see predominantly the vibration that is induced by the planet nearest to it. Typically in a 5-planet gearbox the phasing of the meshes between the planets and the annulus will be arranged so that there is a phase difference to prevent all 5 planets forcing together to induce a torsional movement in the annulus, planet carrier or sun wheel. The effect of this phasing can be that the accelerometer first sees planet 1, exciting vibration at tooth meshing frequency, then planet 2 exciting the same frequency but 72° later in phase, planet 3 later by 144°, etc. After a complete revolution of the planet carrier the vibration has gained or lost a complete cycle and so the observed frequency is not tooth meshing but either 1 or 2 cycles more or less every carrier revolution. The apparent effect is a single 'sideband' but no carrier.

Modulation may cause considerable problems when attempts are made to set permissible levels of vibration in equipment: the effect is to produce sidebands that will contain appreciable power and may deceive an operator or automatic level sensing routine in a computer.

Specification of, say, a permissible vibration level of 10 $\mu$m rms at pump plunger frequency would mean that an irregular pump, modulated at once per revolution, giving 9 $\mu$m at 1/plunger and 6 $\mu$m at the two sideband frequencies, would be passed. This pump would in practice be far more irritating than one with a single line vibration of 11 $\mu$m at 1/plunger.

Altering frequency analysis bandwidths so that one band covers plunger and sideband freqencies will help slightly but it is probably better to return to the original time signal, bandpass filtered if necessary, and take maximum vibration level as the criterion.

## 10.3 Cepstrum

The cepstrum is obtained by taking the frequency analysis of the frequency analysis of a vibration signal. The effect can be seen by taking a rather extreme case of a square wave modulated at 'once per revolution'.

Figure 10.5 shows, to a first order, what the frequency analysis of the signal will look like. Each harmonic will have sidebands, equally

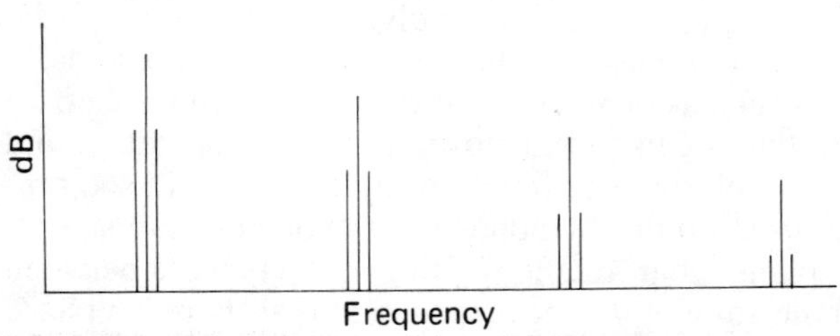


**Figure 10.5** Frequency analysis of modulated square wave

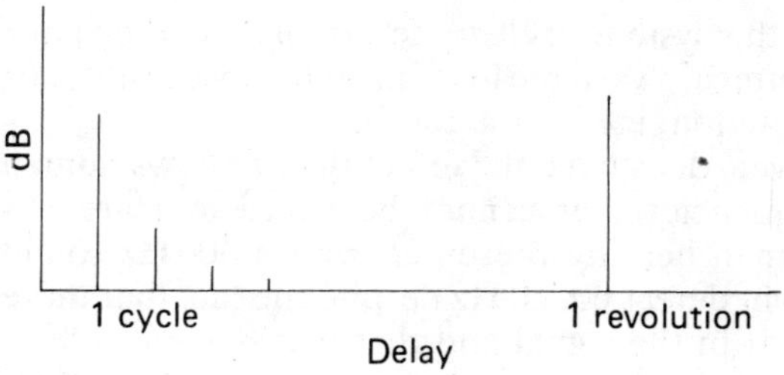


**Figure 10.6** Cepstrum, i.e. frequency analysis of analysis in 10.5

spaced either side, and the effect will be much the same whether frequency or amplitude modulation is used.

Figure 10.6 shows the corresponding cepstrum obtained by taking the frequency analysis of 10.5, and the analysis of pulses spaced at 'once per revolution' will have a strong component corresponding to this spacing. There is no extra information contained in the second analysis but it is presented in a way that accentuates the modulation effects and is less concerned with the 'carrier' frequency.

The fact that there is modulation of a signal shows up as sidebands in a Fourier analysis and thus the cepstrum approach is not likely to give useful information in simple cases, but it may be of use when there are many effects modulating simultaneously and so giving a very complex frequency analysis. In general, the information is not very helpful since, for example, from a normal single mesh gearbox we would expect to see tooth mesh frequency (and harmonics) modulated at once per revolution. Complex systems with several shafts running at different speeds can be investigated using a cepstrum approach but averaging techniques are likely to be more powerful.

Cepstrum is a very useful approach when finite time delays such as those of echoes are involved since a time delay will give a single line at the relevant time (Ref. [1], Section 8.3.1).

## 10.4 Deductions from modulation

The preceding comments have suggested that in practice it is extremely difficult to deduce from a frequency analysis whether modulation is amplitude or frequency or a mixture of the two. This makes it unlikely that much can be inferred from the presence of sidebands as to the likely causes of the modulation, and it is necessary to look at the system itself to determine whether poor gears, eccentric mountings, a variable load, an unbalanced motor or an irregular pumping action, etc., is the cause.

Modulation, however, does have the effect that it allows human beings to 'hear' frequencies that cannot be heard and are not actually present. Human hearing presented with a 500 Hz sound modulated at 20 Hz will detect the 20 Hz despite the fact that there is no 20 Hz component in the signal and that it is not possible to hear a 20 Hz pure tone.

The observation that there is modulation of a note, usually by seeing sidebands, is of relatively little help if the frequency is an expected frequency such as once per revolution, since eccentricity and loads which vary regularly cause such effects. Occasionally there is modulation at an unexpected frequency, giving a clue to problems in the system, but this is rare.

There are other problems associated with modulation deductions, indicated by the diagrams in Figure 10.7. The curve shown in (*a*) can be considered as the multiplication of a steady sine wave by the envelope in (*b*). Whenever two signals are multiplied together the resulting signal has the components corresponding to all the frequency analyses of the two separate signals and all sum and

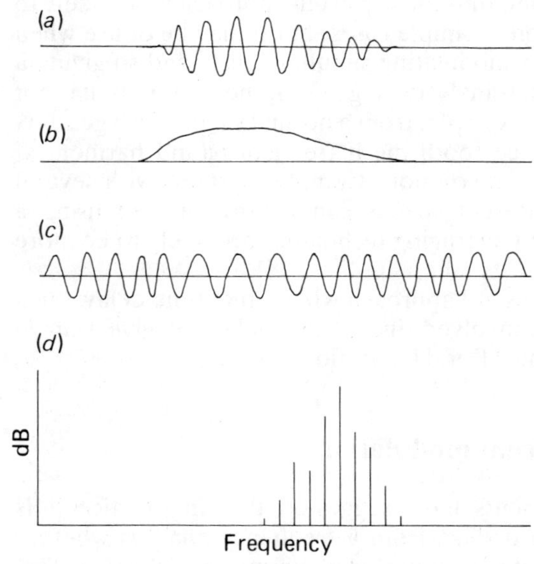


**Figure 10.7** A short burst of amplitude modulated carrier as in (*a*) will give very similar frequency analysis (*d*) to the frequency modulated signal (*c*)

difference frequencies. Hence there will be components in sidebands at many frequencies spaced at multiples of once per revolution, exactly the same as (*c*) which is a pure frequency modulation, and thus both will give the analysis shown in (*d*). Once again, two signals that are very different in character give the same frequency analysis, and so reverse deductions cannot be made.

Modulation of a 'steady' signal thus indicates that something is happening at a particular frequency, but it gives little clue as to what is happening. It does, however, give a broad hint that it may be worth while looking at a signal which has been averaged at a 'once per revolution' corresponding to the modulating frequency.

One technique that can prove useful occurs with very low frequency modulation of a noise such as a gear whine. Modulations of sound at about 1 Hz with speed variation cannot normally be detected as sidebands since they are too near the carrier frequencies, and yet the modulation can be heard. There may be four or five different carrier frequencies and if it is necessary to determine which is being modulated this can be done by continually replaying the recorded sound through a loudspeaker, cutting out each of the carrier frequencies in turn. This technique will work even with short recordings under conditions of varying speed because the human ear seems to be capable of remarkably sophisticated filtering.

## 10.5 Random disturbances

As the essence of modulation is the effect of a regular (usually sine) wave modulation on a 'carrier' such as tooth or plunger frequency, it follows that irregular or intermittent disturbances will not show up well. The converse is that any form of frequency analysis or cepstrum approach cannot be expected to deal with such irregularities and if there is a suspicion that they are occurring different methods must be used.

The type of disturbance that only occurs once, e.g. failure of a system, is an extreme case where a transient recorder is invaluable but where it is possible to catch transients in modulation. In the case of amplitude modulation of a signal the envelope of the signal can be extracted, usually by rectification, and triggering of the transient recorder can then occur on any sudden change of amplitude whether increasing or decreasing. Frequency modulation requires that the signal is first demodulated to remove the carrier, as occurs in a normal FM radio, and the recorder can then trigger on the extracted modulation signal.

Chapter 11

# Correlation and averaging

## 11.1 Introduction

Correlation is a term used both in statistics and in conversation and usually means that if two events or occurences correlate they are somehow connected or agree. In the vibration field the term has a precise definition which departs from the usual meaning; in particular a statistician would not conceive of correlating something with itself.

The definition of cross-correlation is shown diagrammatically in Figure 11.1. Signal 1 is sampled, signal 2 is sampled a time $\tau$ later, the process is repeated for all samples on signal 1 and the product of the two values is averaged over a long period. The value of this average is then plotted against $\tau$. In mathematical terms

$$\frac{1}{2T}\int_{-T}^{T} f_1(t)\, f_2(t+\tau)\,\mathrm{d}t \quad \text{as } T \rightarrow \infty.$$

Autocorrelation is the same process but the samples are both on a single signal.

Three simple cases can be taken to show the effects of correlation on important types of signals, as illustrated in Figure 11.2. Signal ($a$), a sine wave, autocorrelates unchanged ($b$) (as can be checked at the four 90° points); signal ($c$) a train of square pulses gives an autocorrelation function which is a string of triangular pulses ($d$).

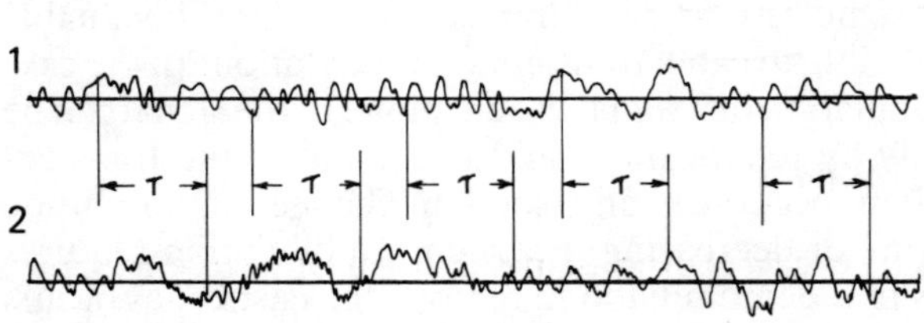


**Figure 11.1** Diagram of correlation. Each sample on trace 1 is multiplied by the sample time $\tau$ later on trace 2

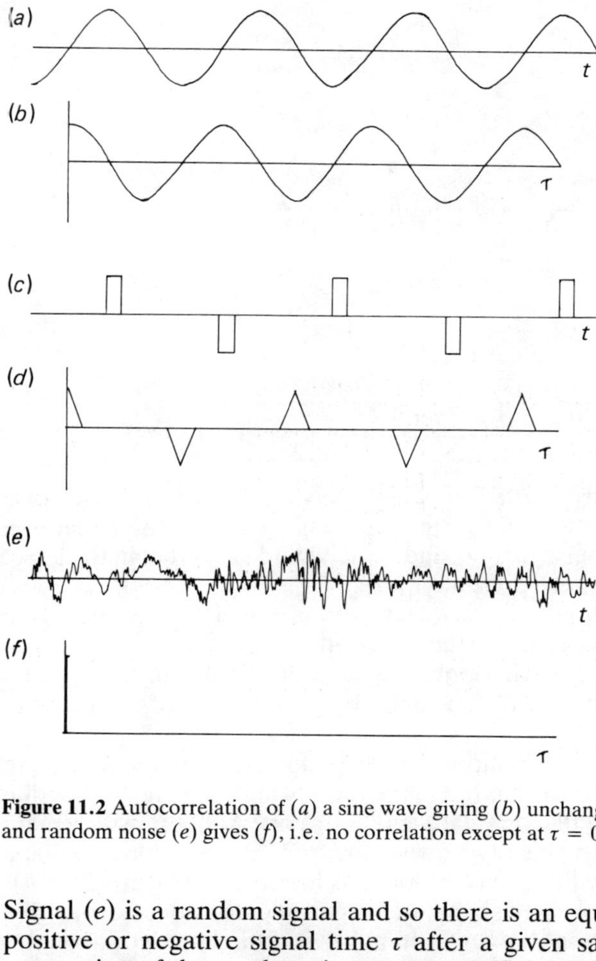


**Figure 11.2** Autocorrelation of (*a*) a sine wave giving (*b*) unchanged, (*c*) gives (*d*), and random noise (*e*) gives (*f*), i.e. no correlation except at $\tau = 0$

Signal (*e*) is a random signal and so there is an equal chance of a positive or negative signal time $\tau$ after a given sample, and the summation of the products is zero over sufficiently long time. The autocorrelation function is zero for all values of $\tau$ except zero; for zero delay the signal is multiplied by itself and averaged giving the familiar mean square value of the signal.

## 11.2 Autocorrelation

The principal use of correlation is in the form of autocorrelation of a single signal when we wish to see whether there is a (small) repetitive signal present in the presence of a large amount of noise. Figure 11.3(*a*) shows a signal of interest, buried in noise as in (*b*).

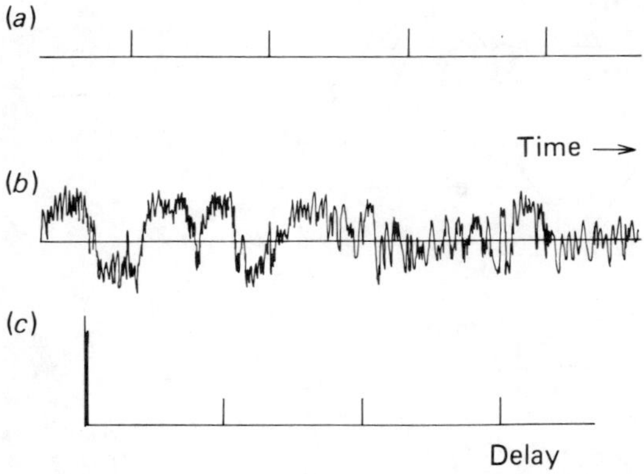


**Figure 11.3** Autocorrelation of signal (*a*) buried in noise (*b*) will give (*c*)

If we merely carry out a frequency analysis, the power in the noise signal at the frequencies of interest may well be greater than the power in the pulse string, and so it would be difficult to detect whether pulses were present or not.

Autocorrelation will, however, give a signal that is large (the mean square value) at $\tau = 0$ and at all other delay times the noise will average to zero, leaving only the autocorrelation function of the pulse string, which will be a similarly spaced string of pulses, as in Figure 11.3(*c*).

This ability of correlation to 'clean-up' the signal and remove random noise effects has led to the term 'whitewashing' to describe the process. If the original signal is a sine wave it will remain unchanged, but a pulse as shown in Figure 11.2(*c*) will have its shape altered and it will be twice as long, as indicated in Figure 11.2(*d*). A pulse of complicated shape, such as an impulsive decay, may give an autocorrelation shape which makes it difficult to work back to the original pulse shape.

The autocorrelation process is thus useful for detecting whether or not a signal is present and obtaining its characteristic time delay or repetition frequency. The disadvantages are that it distorts the shape of the pulse and that, perhaps more important, it requires a very large computational effort. If a test run of 10 000 data points is taken and we wish to calculate the correlation function for a single delay of 73 intervals we must take roughly 10 000 values and multiply each one of these by a value 73 points later. As computers are relatively slow at multiplication and as 10 000 multiplications must be carried out for each of the, perhaps 500, different intervals the computing tends to be slow.

Nevertheless, autocorrelation to detect whether repetition is occurring can be followed by averaging to give clear information.

## 11.3 Correlation

Correlation is sometimes called cross-correlation to distinguish it from autocorrelation. Its characteristics are of course similar to those of autocorrelation, though there is the important difference that at delay $\tau = 0$, cross-correlation does not give the mean square value. Some of the important cases are shown in Figure 11.4. A string of pulses as shown in (*a*), which gives a pulse string delayed $t_1$, as shown in (*b*), will give a correlation as shown in (*c*). This correlation will be unaffected by high noise levels on (*a*) or (*b*), or both, provided that the noise is truly random.

A rather more important case is shown in (*d*) and (*e*) where the original string of impulses (*d*) is erratic and the resulting pulses (*e*) are delayed by $t_1$. The correlation function (*f*) will then have a single pulse at $\tau = t_1$. This idea can then be extended to any wave shape on

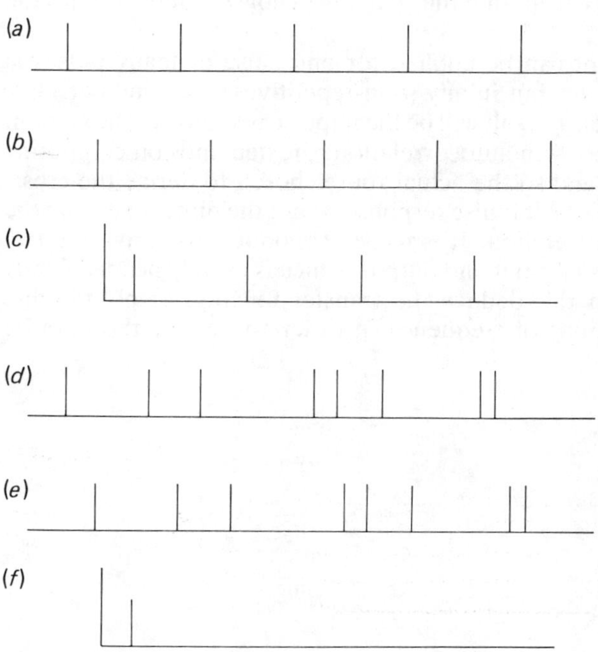


**Figure 11.4** Cross-correlation. (*a*) and (*b*) cross-correlate to (*c*). (*d*) and (*e*) give (*f*)

the 'input' if there is a fixed delay, and whatever the shape of the input vibration, provided it is not repetitive, there will be a single pulse on the correlation at the delay time.

This ability to give a clear cut detection of finite time delays is a major advantage of correlation in fields where it is important to determine transmission time, since this may give a clue to vibration routes or modes of propagation of vibration from one point on a complicated structure to another. Alternatively, cepstrum techniques can be used to achieve delay detection and removal (see Ref. [1], Section 8.3.1).

Another approach to cross-correlation occurs if one signal is the 'input' and the other the 'output' of a system. The process of taking an input sample and an output sample with a time delay $\tau$ and multiplying gives a measure of the connection between output and input since there is a finite averaged result if there is a connection and zero if there is no connection. If we take the simple case of a single unit impulse applied at input, as shown in Figure 11.5(*a*), this will give rise to a response as shown in (*b*), which is the impulse response of the system. Carrying out a correlation will then give a correlation function, since the input is unit pulse, exactly the same as shown in (*b*) except that the '*x*' axis will now be labelled $\tau$ instead of time.

The argument can be applied for one pulse or many pulses at input and so if we put in any (non-repetitive) input and correlate with the output the result will be the impulse response of the system.

In practice, as with autocorrelation, the demands on computing time are large and so the actual route chosen to derive the cross-correlation, i.e. the impulse response, is not the direct one from the definition of correlation. It is more economical to derive the frequency analysis of input and output, which is usually needed in any case, and from this deduce the transfer function amplitude and phase as functions of frequency in order to deduce the impulse response.

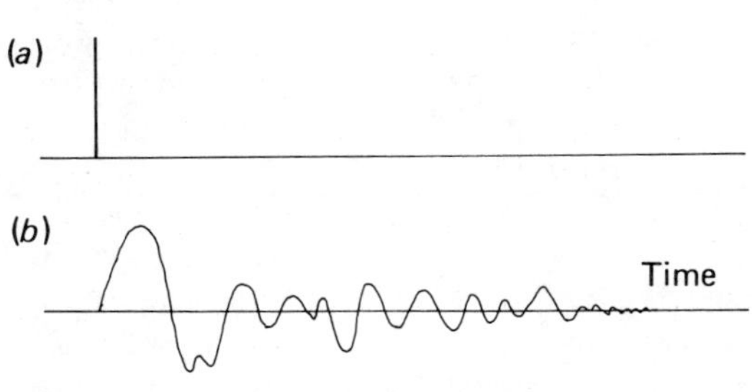


**Figure 11.5** Input to a system as in (*a*) will give the impulse response (*b*)

## 11.4 Time averaging

In many branches of engineering, vibration has a strong repetitive component of interest at a known repetition frequency, usually once per revolution. Reciprocating pumps, gears, rotary printers and IC engines are typical examples where we wish to investigate what is happening at once per revolution (or 2 revolutions) and not at any other frequencies.

Correlation offers a possible solution if we have an accurate once per revolution marker giving unit impulse, since a cross-correlation with the vibration signal will leave only the regular part of the signal. A glance at Figure 11.6(*a*) and (*b*) will show that this approach, though giving the desired result of extracting the exactly repetitive part of the waveform, is very inefficient in computing terms since full frequency analyses of both signals are required and computation will be relatively slow.

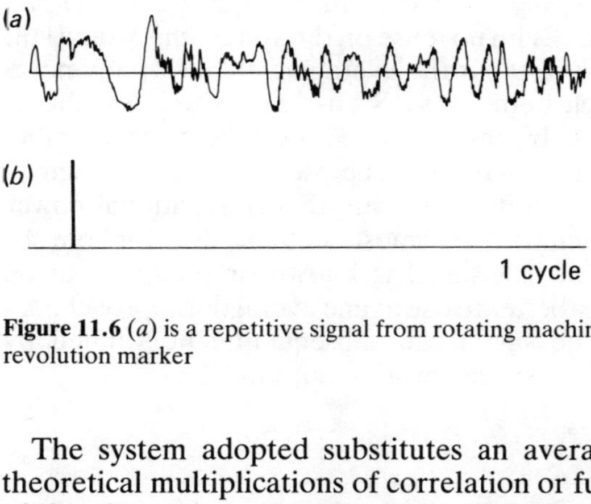


**Figure 11.6** (*a*) is a repetitive signal from rotating machinery and (*b*) is a once per revolution marker

The system adopted substitutes an averaging process for the theoretical multiplications of correlation or full frequency analyses of the alternative route. Typically, 512 samples are taken at each revolution and each sample is simply added in to its respective 'bin' so that the memory requirement is solely for 512 bin totals. Random effects or effects repeating at a different once per revolution interval will average out to zero leaving only the effects from the steady repetitive waveform. A convenient binary number of revolutions, such as 64, can be chosen so that scaling of the resulting average signal is achieved simply by moving the binary total six places to the right. The average obtained is then to exactly the same scale as the input information and is a true signal, not a power signal.

Figure 11.7(*a*) shows a 'raw' vibration signal from an in-line gearbox which has three shafts running at different speeds. Once per revolution pulses from the three shafts allow clear separation of the effects of pitch and once per tooth errors on the three shafts, as shown in Figure 11.7(*b*), (*c*) and (*d*).

Time averaging gives a considerable reduction in computing effort and as it is a very simple process it can operate up to high frequencies without difficulty whilst simultaneously reading out the average signal into an oscilloscope for monitoring purposes. An additional advantage of averaging is that because many samples are taken, the 30 000 mentioned above being a low value, the signal to noise ratio is improved compared with alternative techniques and the final average is very accurate. The average so obtained can be frequency analysed to higher accuracy than the corresponding analysis of the original signal on a limited number of points and the calculation is completed in a shorter time.

It seems somewhat curious that an averaged signal can be considerably more accurate than the signal from which it is derived. This occurs if the original signal was digitised at 8-bit resolution and 64 revolutions would then give a total with an equivalent full scale amplitude of 16 400 bits. With no noise on the system this would still be at 8-bit (0.4%) accuracy but the averaging effects of any noise give an accuracy much better than 8 bits and in practice this is equivalent to around 12-bit resolution. This can be a considerable advantage if a small signal of interest is present with a large signal.

As an approximate rule it can be said that conventional power spectral density techniques are most suitable for random or irregular vibrations but if a signal is known or predicted to be repetitive it is probably better to use a time averaging approach. An accurate averaged time signal can subsequently be studied by frequency analysis (a line spectrum) if required.

## 11.5 Jitter

Averaging is a fast and powerful approach for repetitive signals but inherent in the process is an assumption that the speed of the machinery is steady. Variation in speed gives the effect shown in Figure 11.8 where (*a*) is a signal one revolution and (*b*) is the same signal a few revolutions later when the speed has changed. The summation of the two in (*c*) shows that the speed change degrades the averaged value observed towards the end of the revolution; this effect is referred to as 'jitter'.

Keeping the speed steady sounds an easy solution but if, for example, a gear has 100 teeth, 180° phase shift of a once per tooth component will occur with only 0.5% speed variation. However, if

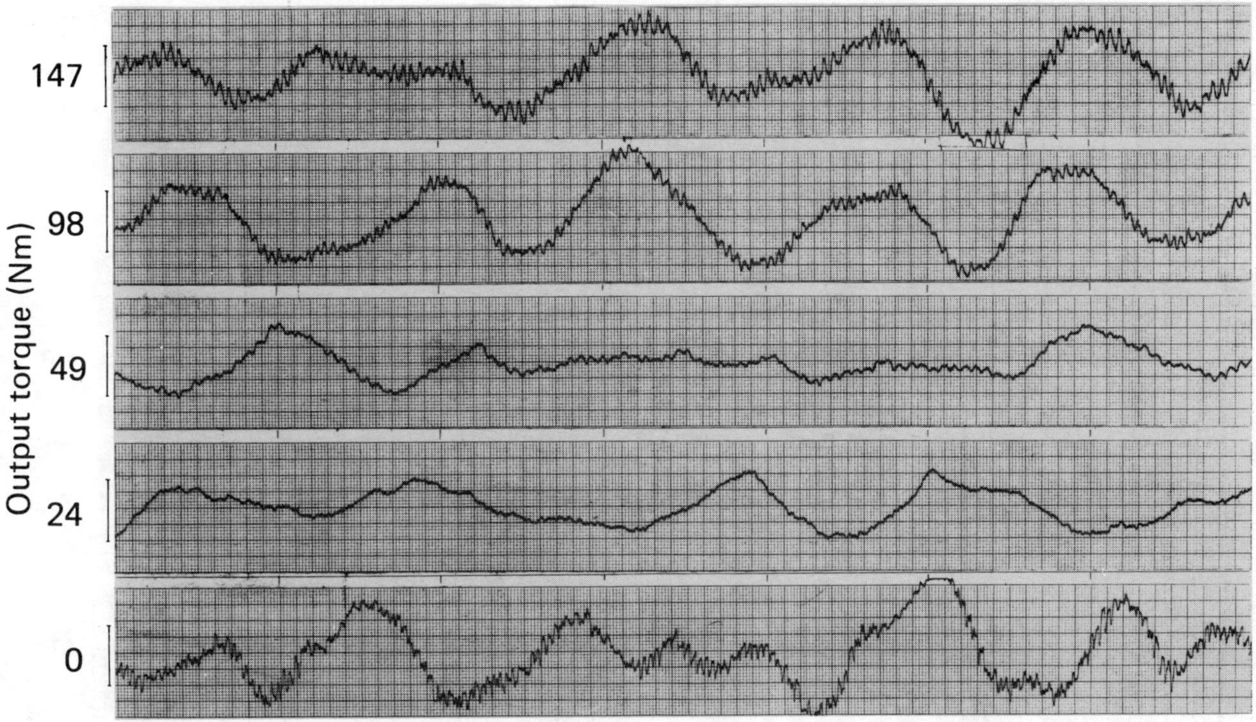


**Figure 11.7** (*a*) Vibration signal from 3 shaft in-line gearbox
(*see also following pages*)

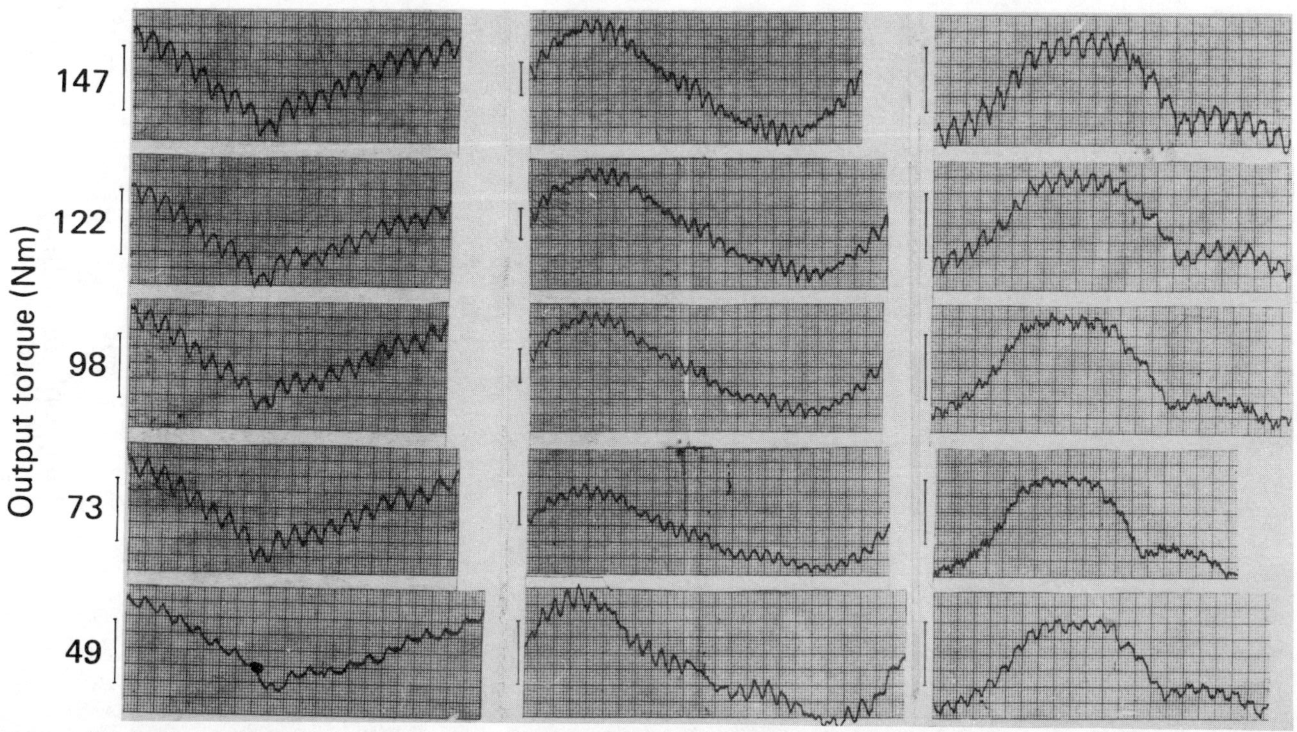
Output torque (Nm)
147
122
98
73
49

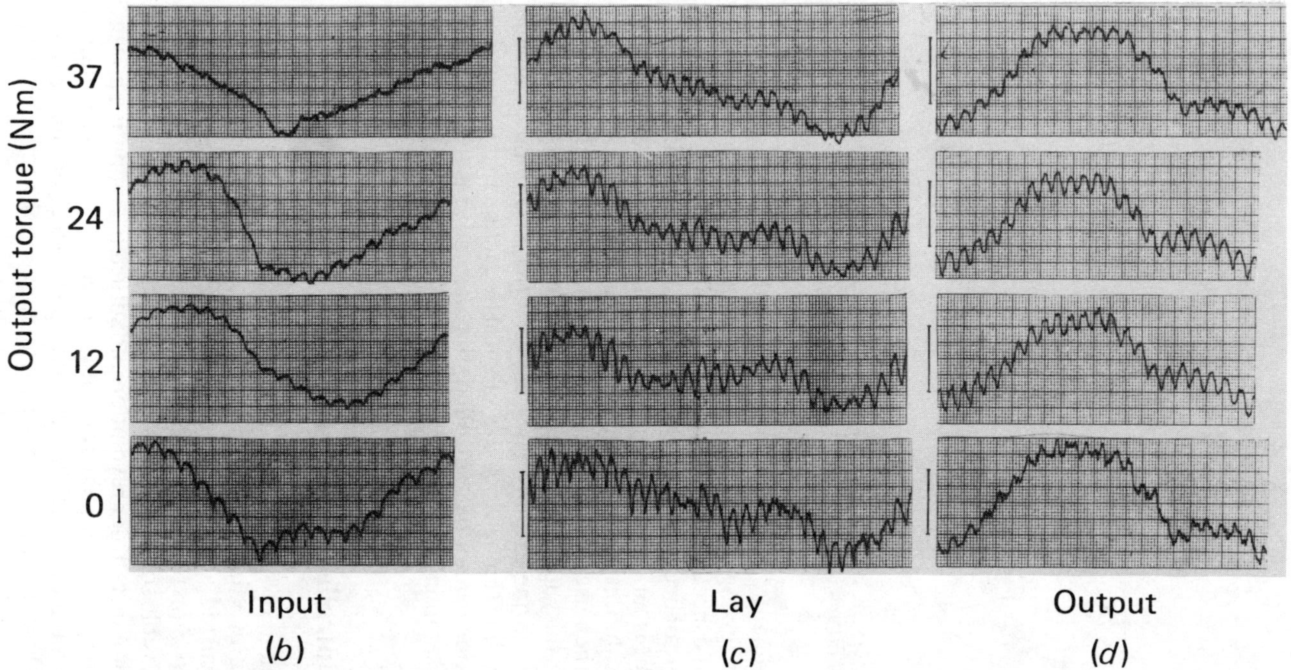


**Figure 11.7** (*cont.*) Vibration signal from 3 shaft in-line gearbox. (*b*) Averaged signal from input shaft. (*c*) Averaged signal from layshaft. (*d*) Averaged signal from output shaft

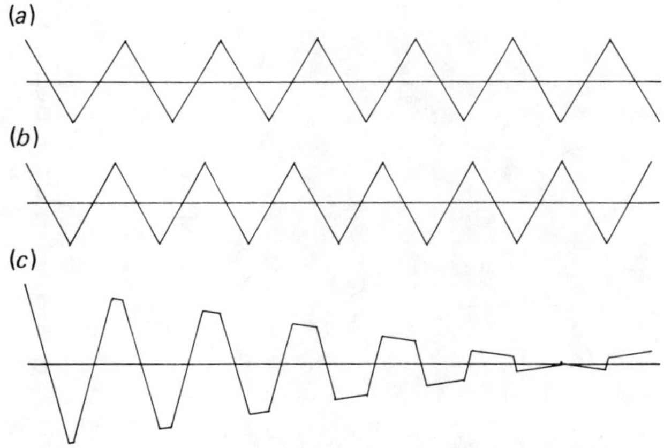


**Figure 11.8** (*a*) Signal from one revolution. (*b*) Signal from later revolution. (*c*) Sum (average) of (*a*) and (*b*)

the component of interest is the third harmonic of once per tooth frequency, giving noise, the speed would have to be held to 0.17% to prevent cancellation of the third harmonic content.

This sort of accuracy might be attainable on a specially designed test rig but is unlikely to be achieved on a practical installation. A possible solution is to average only part of a revolution, say one-quarter, then use a timing pulse one-quarter of a revolution later, etc., but this approach is relatively slow and clumsy. As the problems are due to a mismatch between the revolution speed and the controlling timing clock it is a logical deduction that if the speed of revolution cannot be controlled, the alternative is to control the timing of sampling.

## 11.6 Controlled sampling

The problem of 'jitter' arises because a vibration at 100 times per revolution may incur error in half a cycle and so a timed sample which occurred at tooth 99 on the first revolution might occur at position 98.5 on the next revolution, and cancel.

One possible solution to this speed variation can be achieved with a high sampling rate per revolution and sufficient computing power. The samples for a complete revolution are stored until a revolution has been completed and the next once per revolution timing pulse arrives. The correct position for each sample in the revolution can then be calculated on the assumption that the speed, although varying from revolution to revolution, is consistent within a revolution. Needless to say, this approach involves taking many more initial data points per revolution with corresponding high data rates, then completion of position calculations, and finally storage of the chosen points in the relevant 'bins' before the next revolution has been completed.

The solution usually adopted is much faster and more economical; it consists of controlling sampling by pulses generated from a timing wheel attached to the shaft: 128 or 256 holes are drilled round the periphery, and a phase locked loop driving a voltage controlled oscillator can multiply the frequency up to perhaps 1024 times per revolution, with each pulse exactly locked to an angular position regardless of speed. The sampling interface on the analysis system is triggered by these pulses instead of the normal timing crystal derived pulses. With this system the sampling is consistently at the same set of points at each revolution regardless of speed variation within a revolution or between revolutions.

Most applications of averaging are not as critical as the '100-tooth' case and thus a simple once per revolution pulse is sufficient here.

## 11.7 Average or variation

It is tempting to assume that averaging techniques should only be used when it is the steady repetitive signal that is of interest. This is likely to be the case for many types of vibration such as gears or pump plungers. Experimenters should, however, bear in mind that occasionally it is not the normal 'average' level of the response that is important from an engineering point of view but, rather, the variation from the normal.

A case might occur where one cylinder in an IC engine misfires intermittently for perhaps 1 cycle in 20 or a plunger in a reciprocating pump sticks intermittently due to dirt in the oil. The signal of interest is then not the average but variation from the average and this is displayed most clearly if the average is generated and subtracted from each cycle of vibration so that the observed trace emphasises the differences from the 'norm'.

This facility to view differences from the 'steady state' is often available on frequency analysis equipment but is less useful for pinpointing exactly where in a cycle the differences occur, since the information is in the frequency domain.

Chapter 12

# Transfer functions

## 12.1 Time or frequency

The concept of a transfer function or operator is fundamental to the study of the majority of systems in control or vibration. Theoretical estimates may suggest the 'block diagram' type of description, as shown in Figure 12.1, but there are often complexities or indeterminable quantities such as damping; practical measurement is then required.

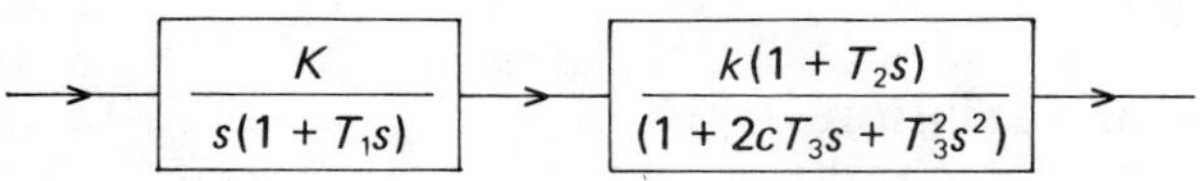


**Figure 12.1** Typical theoretical block diagram of a system

A relationship between outout and input can only occur if there is external excitation, i.e. if it is not a self-excited vibration. The relationship is unique and hence useful only if the system is essentially linear so that the principle of superposition can be invoked. Regardless of whether we describe the system in frequency response terms or time response terms, the assumption of linearity is made so that we can state that the effects of inputs applied separately will sum to give the effect when the inputs are applied together.

In general, there will be a varying input and a corresponding output and it makes no fundamental difference whether we regard the input as an agglomeration of impulses, as shown in Figure 12.2 or, perhaps with a more regular waveform as shown in Figure 12.3, regard it as a collection of superposed sine waves. The natural tendency is to think of impulsive happenings in terms of impulses and regular waves in terms of frequency analyses but the mathematics draws no distinction and so will 'regard' any vibration as either, equally happily.

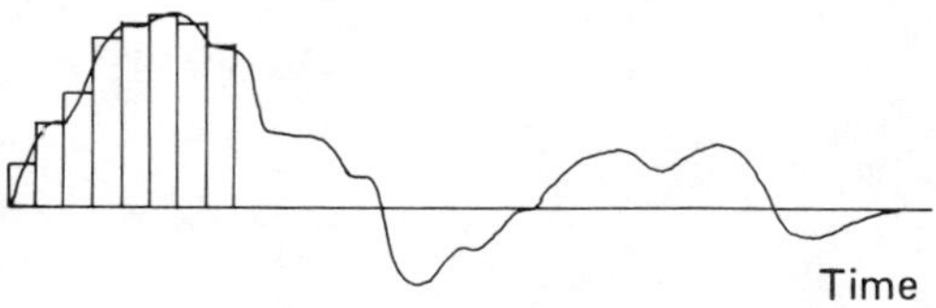


**Figure 12.2** Input waveform split up into a series of impulses

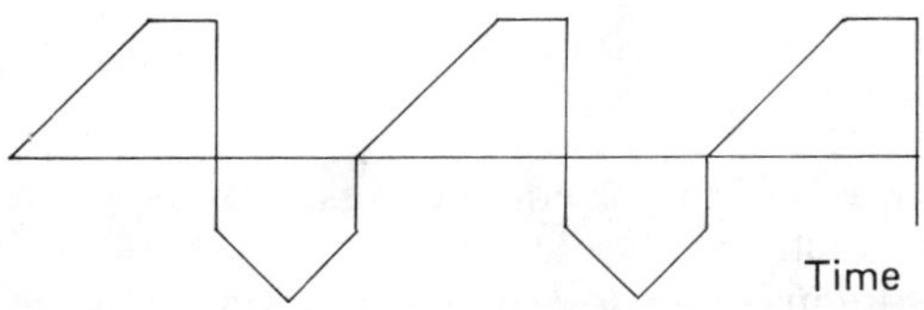


**Figure 12.3** Regular waveform which will give a Fourier series of harmonics

Just as we can specify the input vibrations just as precisely as collections of pulses or sine waves, the transfer function between input and output can be specified as that resulting from an impulse or that resulting from a sine wave. The 'time domain' response and the 'frequency domain' response are inevitably connected by a Fourier transform in mathematical terms. Conceptually, it is equally easy to say that a pure impulse consists of equal powers of all possible frequencies (coincident in phase) and so the 'impulse response' of a system can be obtained by adding the effects of all the 'frequency responses'. Conversely, an input sine wave can be synthesised by a series of impulses and so the 'frequency response' of a system can be obtained by adding the effects of a series of impulses. This latter approach is termed 'convolution'; any shape of input 'pulse string' is multiplied by the impulse response and the results are added, with due regard to time shifts, to give the output.

The final transfer function can be displayed either as a frequency response, Figure 12.4, or as an impulse response, Figure 12.5, and there tends to be much irrelevant argument as to which is 'better'. As so often occurs there is no 'better' system since the choice depends purely on what the customer wishes to see. At one extreme the effectiveness of a point-to-point positioning system for a machine tool is most clearly shown in the time domain since parameters such as overshoot, time to within 20 μm of position, etc., are directly visible. A 'time domain' impulse response would, however, be a very poor way of indicating the effects of a steady excitation such as imbalance in the neighbourhood of a very low

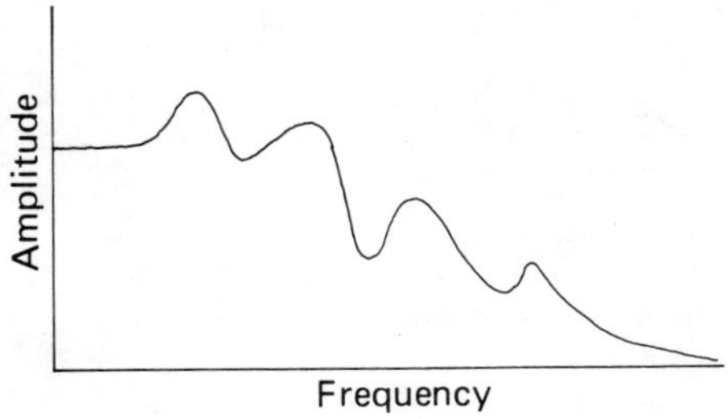


**Figure 12.4** Typical frequency response of system

damped whirling resonance and thus frequency response is much better in this case.

It is very common for the 'frequency response' of a system to be given or displayed with only the amplitudes shown and not the phase. This is especially dangerous in control work since phase is so important, but it is advisable to display phase in structural vibration testing of machines since a sudden change of phase can indicate trouble due to play or backlash long before there is an appreciable effect on amplitude of response.

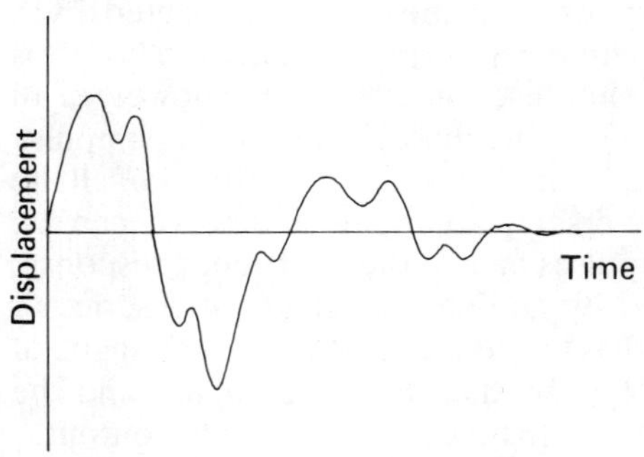


**Figure 12.5** Impulse response of system

## 12.2 Coherence

If we have a single input, single output system we can deduce a relationship between the input and output. This, however, does not necessarily tell us the true transfer function because the output measurement may have large components due to noise or non-linearities in the system. To obtain a valid transfer function the cross-spectrum density is first obtained by multiplying input values (at a given frequency) by output values (at the same frequency) and averaging. The transfer function is given by the ratio of the cross-spectrum density to the input spectrum density.

We can then take the input power, multiply by the 'transfer function' that we have deduced and hence predict what the output power should be. Agreement between the measured output power and that predicted by the product of input power and transfer function suggests that all is well. If the measured value is greater than the predicted value, then extra power has entered the system from noise or there are non-linearities and we cannot depend on the transfer function.

Coherence is a measure of the ratio of predicted power to actual output power. One extreme case occurs when coherence is 1.0, the maximum possible value, in which case we can with confidence say that input and output are linked. A coherence value of 0 means that all the output power at that frequency is due to noise or unrelated effects in the system and so the transfer function deduced is not valid and should be ignored.

There is one particular case where coherence deductions are not valid due to an oddity in the mathematics of sampling. This occurs when a relatively small number of samples (1024 typically) is taken and analysed using an FFT approach and tends to give a coherence of 1.0 even when noise is present. The effect is discussed in specialist books [1] and the effects can be avoided by smoothing spectral estimates.

The majority of twin channel vibration analysers now have a readout of coherence and it is thus a simple matter to check the validity of transfer functions by glancing at the coherence function. Where the signals are large, usually near resonances, the coherence will be 1.0, but where signals are small and disappear into noise, usually near important anti-resonances, the coherence will be 0.

A coherence of 1 confirms that 'input' and 'output' are linked and there is a strong temptation to say that if two measurements are made of, say, force and vibration, a coherence of 1 'proves' that the vibration is a result of the force. This is not a valid deduction since, although vibration may be a result of force, force may be a result of vibration or both may be a result of another unnoticed input. As an example, measurement of vibration at one bearing housing on a gearbox can give a coherence of 1 when checked against force transmitted through the antivibration mounts for the whole gearbox. This does not 'prove' that the vibration results from forces through the mounts, or the reverse. In reality, both force and vibration result from transmission error (relative displacement) at the gear mesh.

A major use of coherence is to show there is a fault in the system. In one case a mechanical connector had not been correctly tightened and so slip was occurring in the connector at high vibration levels. The slip gave non-linearities and thereby reduced the coherence between the vibrations measured on either side of the connector.

Another case occurred due to faulty slip-rings transmitting vibration information from a rotating system. Under some conditions the slip rings resonated and lost contact intermittently, and so reduction of coherence gave a clear indication of the critical areas for slip ring operation.

## 12.3 Transmissibility

The basic ideas of an 'input' to a system being operated on by a 'transfer function' to give an output are very standard and are relevant to many engineering systems but can be very deceptive when applied to the measurement of the performance of anti-vibration mountings in a practical installation.

If the system were as shown in Figure 12.6, with input force **F**, and output as the force transmitted, **f**, to the (rigid) floor, then the concept of transmissibility can be used; quoting *f*/*F* is a good measure of the performance of an isolator. Problems arise when the support is not rigid. A typical vehicle suspension may have three routes by which vibration may travel through isolators into the body, as sketched in Figure 12.7. As it is difficult to measure force directly, accelerometers may be used on either side of the isolators to measure spring vibration and body vibration. Taking the ratio of the two amplitudes on either side of the isolator is *not* a valid

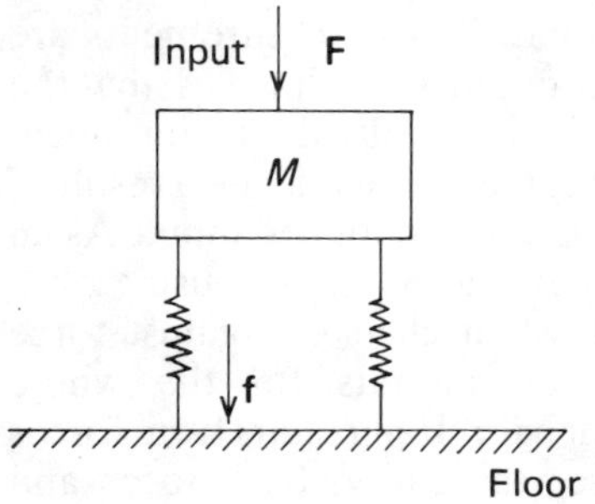


**Figure 12.6** Simple vibrating system with rigid floor

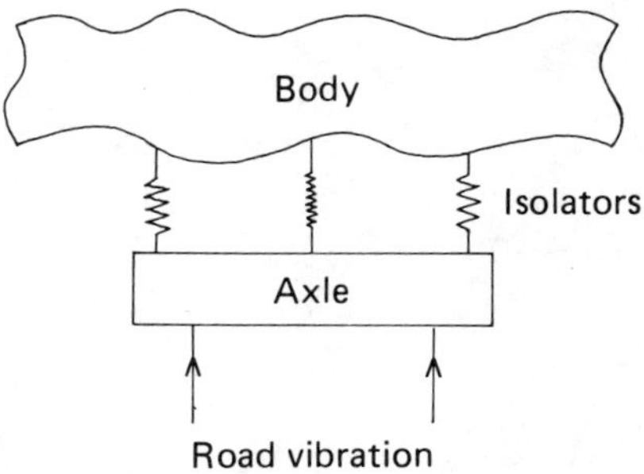


**Figure 12.7** Realistic system with body of variable local stiffnesses

measure of the effectiveness of the unit since the body vibration is influenced not only by the force transmitted but also by the local body stiffness. A high local body stiffness, possibly near an anti-resonance, will give a good ratio of vibration attenuation across the isolator and the impression will thus be that the isolator is 'good'. In reality the power transmitted into the body will be high and will travel to the absorbing resonance elsewhere (which is causing the local antiresonance) and will give high noise levels. Conversely, high local amplitudes associated with a local body response will not necessarily transmit much power into the vehicle.

When vibration can travel via several routes, as in this case, the only reliable method is to measure the force transmitted at each point and multiply by velocity to give the power being transmitted into the vehicle for each route separately. Measurement of the force is simplest if a transducer is fitted between body and isolator but if this is not possible the isolator should be calibrated in a test rig so that its force–deflection characteristics are known exactly and, *in situ*, the force can be deduced from the relative vibration between the ends of the isolator; it is of course important to take note of the relative phase. Calibration should be carried out at the correct average load and with the amplitudes and frequencies encountered in practice because characteristics are often non-linear. If large movements of the whole isolator occur on the vehicle it may be necessary to mass compensate the results.

## 12.4 Identifying resonances

A complicated system with a large number of resonances will have its resonances so close together that their effects interfere; it is no longer possible to use the single degree of freedom approach of maximum amplitude or, better, 90° phase, to identify the natural

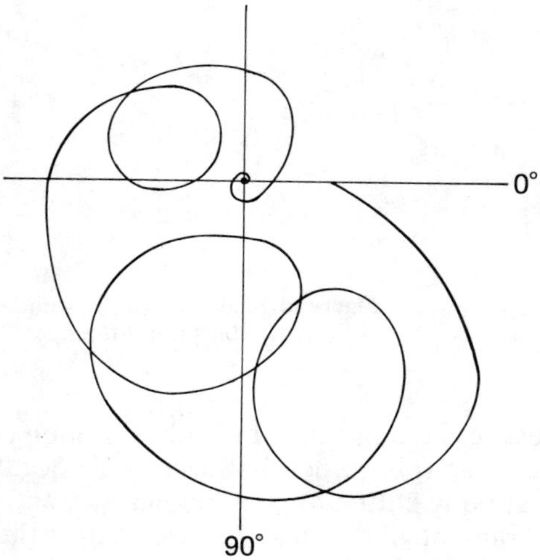


**Figure 12.8** Response plot of system in polar form

frequency and then half-power points to decide the bandwidth and hence the damping. A plot as shown in Figure 12.4 gives no clear idea of the characteristics of each resonance.

The polar, vector or Nyquist plot as shown in Figure 12.8 will consist of a number of loops, and Kennedy and Pancu [2] developed a method to separate out the effects of each resonance, i.e. each loop. They first made the assumption that the damping was hysteretic, not viscous; this turns the normal single degree of

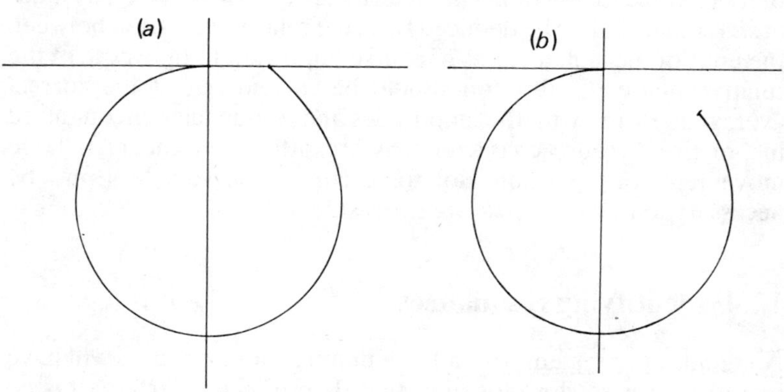


**Figure 12.9** Single degree of freedom system response (*a*) with viscous damping, and (*b*) with hysteretic damping

freedom polar plot (Figure 12.9(*a*)) into a circle with a small section missing, as in Figure 12.9(*b*). This assumption, as it happens, is usually more accurate than the normal viscous damping assumption, which is chosen solely to give a linear equation for easy solution. In practice the polar plot loops are remarkably good approximations to circles and so a given resonance can be identified by locating its radius to determine the amplitude of that resonance and the 3 o'clock and 9 o'clock points that correspond to the familiar half-power points. If the equation is written as $m\ddot{\mathbf{x}} + \lambda\mathbf{x}/\omega + k\mathbf{x} = \mathbf{F}\cos\omega t$, the circle radius is $\mathbf{F}/2\lambda$ and at the 'half-power' points $\omega_o/\Delta\omega = Q = k/\lambda$.

Location of the circles can be performed very effectively by eye but when the process is done by computer it is more likely that identification will be by a curve fitting technique, adjusting parameters on conventional viscous single degree of freedom systems to give minimum residual differences from the experimental curve.

The techniques for using computers to disentangle and identify neighbouring resonances are now very comprehensive and detailed. If a really complicated problem arises the approaches discussed by Ewins [3] should be investigated. Occasionally, however, the use of some ingenuity may help if it is possible to locate, even roughly, the positions of nodes of the interfering vibrations and excite these positions in order to obtain a relatively pure resonance.

## 12.5 Mode shapes

Associated with each resonance is a characteristic mode shape; in mathematical terms each eigenvalue ($\omega^2$) of the system has a corresponding eigenvector. For a simple system, such as three masses on a beam we can specify the mode shapes simply, as indicated in Figure 12.10; the mode shapes may be drawn or given as (1, 1.4, 1) (1, 0, −1) and (1, −1.4, 1). In this case the mode shapes are as expected or calculated and there is no useful information in them.

The practical case of a large machine tool is rather different because from the workpiece markings the troublesome resonance frequency will already be known, but the factor of interest is the shape in which the machine is bending. Experimentally, a vibrator will be set at the cutting tool position to give resonance, and a pick-up will be moved around the machine to give the vibrating shape at resonance; phase is checked to give positive or negative displacement. As some 100 points may be taken, an eigenvector would be a

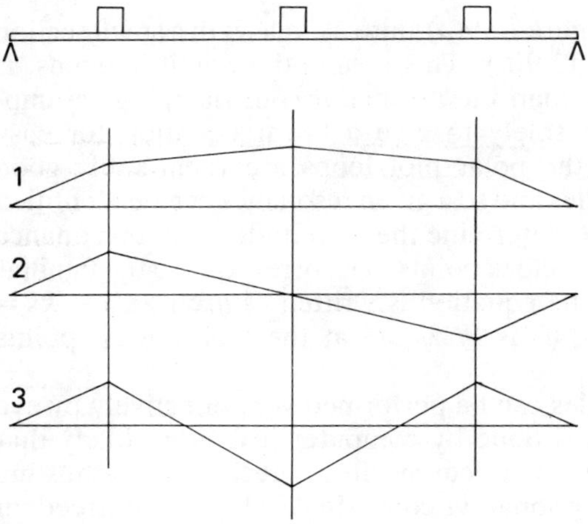


**Figure 12.10** Three degree of freedom system of three masses on a light beam showing the three mode shapes

clumsy way of showing the information and it is thus better to plot it as shown in Figure 12.11.

A discontinuity in the mode shape shows play or backlash, usually at a bearing or slideway; in general, phase of the signal will also change significantly when this happens. More commonly the feature of interest is a discontinuity in slope; this shows that there is a 'weak' spot in the machine which is almost acting as a hinge and is thus giving reduced stiffnesses and natural frequencies. Any region of high curvature in the mode shape should be investigated in more detail to determine how the metal is distorting.

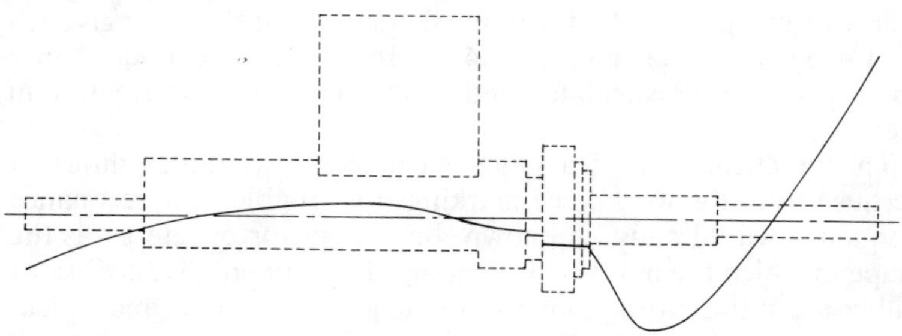

**Figure 12.11** Practical mode shape of large machine tool (full line shows deflection from centre-line)

## 12.6 Modal analysis

The method described above gives mode shapes but is relatively slow and unfashionable. The same information can be obtained by attaching a large number of accelerometers along the machine and exciting either by vibrator or hammer at the cutting tool. Application of white noise or an impulse allows all frequencies to be dealt with simultaneously and, given sufficient accelerometers, the mode shapes can be identified. When resonances are in very close proximity the computer can use the phase information to separate out the responses of the different modes. Each mode shape can then be displayed separately or printed out to give a convenient permanent record. (But see Section 16.3.) Such an approach is popular because it involves the use of relatively fast operating, high computing power, and hence an expensive and impressive installation that is very good for sales purposes.

A slight variation on this approach is to economise on the work of attaching many accelerometers by invoking the reciprocal theorem. One accelerometer only is attached at the cutting tool position and an impulse hammer is taken round the machine. The position of each impulse must then be fed to the computer if it is to present mode shapes.

As mentioned in Section 12.4 the response is complicated when resonances are in sufficient proximity to interact, and a curve fitting technique can be used to advantage. However, high accuracy in mode identification can best be achieved by using many vibrators placed and programmed to suppress the other natural modes leaving one mode isolated for detailed investigation. This approach is so expensive that it is usually only used for aerospace work.

Chapter 13

# Signal filtering and improvement

## 13.1 Uses of filtering

The most common requirement for filters is to remove high frequency 'noise' from a system. This is the standard low pass filter that is intended to pass signals up to a cutoff frequency without distortion and also to remove higher frequency signals.

Pure analogue processing of signals does not usually require sophisticated filters. 'Noise' in a typical machine set up may be due to electrical noise in the measurement system or to carrier frequency rectification effects, but it is more likely to be due to impacts within the machine, producing small high frequency vibrations that give high accelerations but that are unimportant. Such noise is of higher acceleration amplitude than the signal of interest but it is at a high frequency so that if a cutoff filter is set to, say, twice the signal frequency of interest, any noise at 20 times the frequency is greatly reduced.

The more modern digital systems involve sampling the signal and, with this approach, filters are necessary to reduce aliasing effects, as discussed in Section 7.7. The filters required for this application must have appreciably higher performance since a frequency range of little more than 4 : 1 may be available between signal frequency and corresponding alias frequency. As it is advisable to reduce an alias frequency amplitude to, say, less than 1% of the signal frequency amplitude, a filter should achieve 100 : 1 reduction within a 4 : 1 frequency range and so have a minimum 20 dB/octave cutoff. Even this would give some signal distortion.

An even more extreme requirement for a 'low pass' or 'high cut' filter can occur in the case of transmission error measurement systems. Under some circumstances the signal of interest may be at 20 Hz and be of amplitude 0.1 V, while the 'carrier' signal from the phase comparison on which the system works may have a frequency of 40 Hz but an amplitude of 10 V. Within an octave we thus need a reduction of at least 1000 : 1 so a filter with at least 60 dB/octave cutoff is required.

High pass filters are relatively rarely used as separate units, though the conventional AC coupling is a particular case where all the vibration part of a signal must be passed but where the DC component is not required. The cutoff frequency may be about 0.5 Hz, sufficiently low for most purposes.

Bandpass filters consist of high pass and low pass filters connected in series and set so that a band of frequency is passed. They are of most use when investigating a particular resonance in a system if excitation of the resonance is erratic or transient.

An alternative to filtering the signal from an accelerometer is to integrate the signal to velocity or even to double integrate to displacement. As amplitudes integrate proportional to the inverse

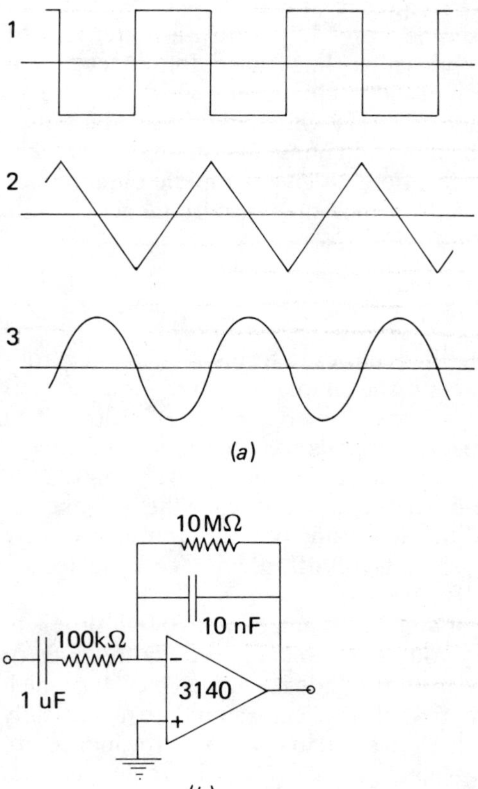


**Figure 13.1** (*a*) Effect of integration of acceleration square wave 1 to triangular velocity 2 and to displacement 3 which is parabolic. (*b*) Typical integration circuit with 1 ms integration time

of the frequency, noise at 20 times signal frequency is reduced at each integration by a factor of 20 relative to the signal. A striking example of the smoothing effects of integration is shown in Figure 13.1(*a*) where a square wave with high harmonic content integrates to a triangle and then to a series of parabolae which are indistinguishable from a sine wave. One precaution that is necessary when integrating is to have a DC block, i.e. a high pass in the system, otherwise DC will integrate to produce saturation of the system. The circuit normally used is shown in Figure 13.1(*b*) with typical values to give a low frequency cutoff at about 1.6 Hz and a break-even or 1 : 1 amplitude frequency of 160 Hz.

Notch filters are used occasionally to remove unwanted noise signals usually at 50 Hz, but they also remove the signal in the neighbourhood of the set frequency and so are of limited use. Digital line elimination (see Section 13.7) is now a better way of achieving freedom from particular line noise frequencies. An alternative 'old fashioned' method of eliminating 50 Hz is to use an operational amplifier to add in 50 Hz which neutralises the noise; signals are derived from the mains, in phase and in quadrature and sufficient is added to the vibrations to give complete elimination. Testing can then occur reliably even at the noise frequency.

## 13.2 Filter order

It is customary to describe filters either in terms of 'order' or 'roll-off' rate. A first order system is the simple R–C filter and the gain characteristic on log scales is as shown in Figure 13.2(*a*). The frequency at which the capacitor impedance equals the resistance is known as the break frequency, 45° point, half-power point, 3 dB point, cutoff frequency, etc. Above this frequency the influence of the capacitor controls and the amplitude of the output is inversely proportional to frequency so that 'roll-off rate' is 20 dB per decade or, more conveniently, 6 dB per octave.

A second order system obained by connecting two first orders in series does not give a very good performance; see Figure 13.2(*b*). Near the 'break frequency' the amplitude is reduced by 6 dB, which is too much, and as two first orders correspond to a critically damped vibrating system the phase changes before resonance are high (at resonance the phase change is 90°). Considerable improvement is obtained by reducing the 'damping' from critical; this gives an improvement in signal amplitude near resonance and reduces phase effects. The exact amount of damping selected depends on whether it is considered more important to maintain phase changes

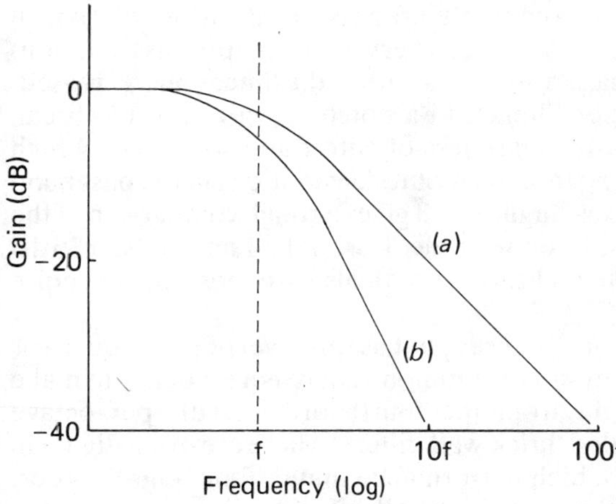


**Figure 13.2** Filter characteristics. (*a*) is a simple lag circuit and (*b*) is two similar lags in series

to a minimum, in which case low damping is required, or whether amplitude is critical, requiring rather more damping of about two-thirds of critical. The roll-off rate associated with a second order corresponds to an inverse square law at 12 dB per octave; a typical 4 : 1 ratio between signal and noise frequencies gives 24 dB reduction, which is sufficient if noise is already significantly smaller than the signal.

Third order systems with a phase change of 3 × 45, i.e. 135°, at resonance and a corresponding roll off rate of 18 dB per octave are virtually unused despite their apparent convenience of giving nearly a 'rule of thumb' performance of reduction to 1% over a 2 octave range. Part of the reason for this is that in practice the filter is set somewhat higher than the highest signal frequency in order to reduce distortion. For a signal frequency ranging up to 500 Hz, the filter frequency might be set to 700 Hz and thus half an octave is 'lost', leaving only 27 dB reduction for 2000 Hz noise. This combines with design problems to ensure that in practice a fourth order system is generally used.

Fourth order systems are very common and give a 24 dB per octave performance with 180° phase lag at the break frequency. Detailed design is complicated to achieve compromises between constancy of amplitude and phase distortion reduction; two second order systems in series can be used with adjustment of dampings and frequencies as required. There is no 'best' solution and various designs are currently in use.

The standard, most commonly used, is the Butterworth, which gives the best flat response and is a very good compromise solution. When linearity of phase response is critical the 'linear phase' is used. Chebychev and Cauer filters add a notch effect above the break frequency to emphasise sharpness of cutoff and so give very high roll-off rates at the expense of amplitude variations in the pass band and at high frequencies. Figure 13.3 gives a rough comparison of the most common types. Even with the 'loss' of half an octave a 36 dB reduction is available and this order deals with most normal noise problems.

Higher order systems are used, but as this level of performance is not often needed, they are uncommon. There is a tendency to make the transition straight from the fourth order 24 dB per octave systems to the so-called 'brick wall' filters. These are often digital in operation and have a high performance that is about eighth order, but are Cauer and thus give high roll off, a typical figure quoted being 135 dB/octave. Such filters are inevitably expensive and are used when there is a very narrow frequency margin between signal and noise, or when the 'noise' is a carrier wave that is much larger than the signal. It is possible to obtain an 80 dB reduction over a 1.5 to 1 frequency range.

It is sometimes worth remembering that tape recording can act as a very effective filter, albeit at a set of rather fixed frequencies. This is due to the characteristics of the tape when direct recording, or the

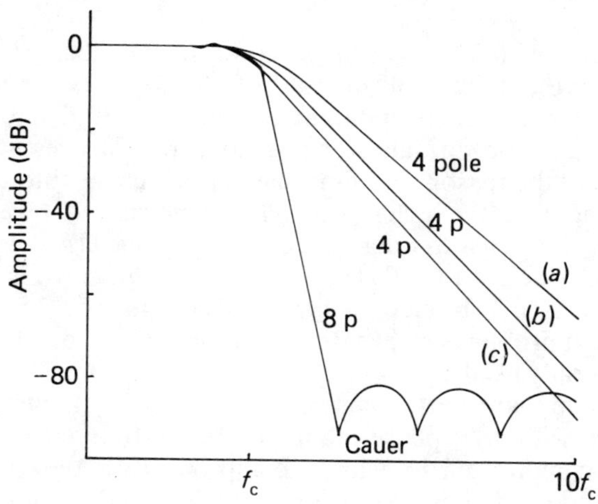


**Figure 13.3** Typical filter responses. (*a*) is a linear phase filter, (*b*) is the standard Butterworth, (*c*) is the Chebychev, all 4-pole filters and (*d*) is an 8-pole 'brick wall' Cauer filter

system filters with frequency modulation. Careful selection of tape speed not only economises on tape but also reduces filter requirements.

## 13.3 Compromises in selection

At first sight it would seem simple to select the highest possible order system that was available and use this for all purposes, finance permitting. In practice the fundamental laws of analysis seem to say that no advantage ever comes without a corresponding problem.

The basic choice between, say, amplitude or phase performance of a fourth order filter can be relatively simple once the engineering requirements of the problem are known. In general, amplitude is more important since the objective is to remove high frequency noise and phase problems can be overcome simply by putting all channels of information through identical filters so that phase delays are matched; this is, of course, wasteful since channels are filtered unnecessarily.

Use of higher order filters gives better roll off and noise reduction with the corresponding penalties of higher phase distortions as the frequency approaches the break frequency and, if the Cauer types are used, additional slight amplitude fluctuations with frequency. Another problem can occur with 'ringing'; this is the response observed when an impulse or step function excites the filter. Because a Butterworth filter behaves in a similar manner to a two

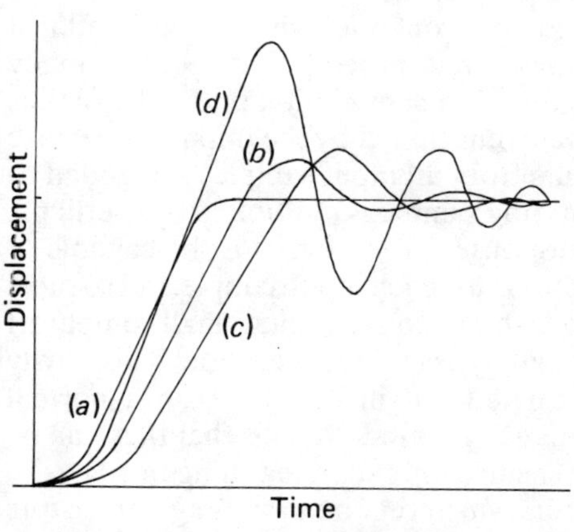


**Figure 13.4** Responses to step input change (*a*) linear phase, (*b*) Butterworth, (*c*) Chebychev and (*d*) Cauer 8-pole

degree of freedom system with less than critical damping, the response is as shown in Figure 13.4. High damping designs such as the linear phase give no ringing, but designs such as the Cauer give a very large overshoot and as the effective damping is low there are several cycles of oscillation at a frequency corresponding to the 'break' frequency. A step or impulse function is relatively uncommon in normal vibration work but may occur with pseudo-random excitation and can, of course, occur with impulsive inputs.

Care is needed whenever a step change occurs at input to a standard filter. A square wave of ±5 V at input to a filter with specified input of ±7 V would seem permissible. The sudden 10 V transient at each transition tends to ring, i.e. overshoot and exceed the linear range of the input to give incorrect results when accurate phase measurement is attempted. To protect the filter input from the sudden step change a preliminary, R–C, first order filter must be used. This 'pre-filter' should be fitted at the output from the previous stage, otherwise the theoretically infinite rate of change of voltage applied to the finite capacitance of the connecting lead will attempt to draw high current and saturate the previous stage output.

## 13.4 Analogue or digital

It is customary to use analogue low pass filters at input to a digital analysis system and, as it is possible to use either digital or analogue filtration, it could be asked why the system does not remain completely digital. Alternatively in some set ups a digital filter precedes an analogue device, such as a pen recorder.

The choice is essentially one of economics with no hard and fast rules. Digital filtering is, however, a process that requires many calculation steps. The procedure is a convolution integral and thus each sample of input is digitally multiplied by the impulse response of the filter and all the results from all input samples are added to give the output. As each entering sample is multiplied by, perhaps, 100 points on the impulse response the system must be capable of 100 multiplications and additions for each input sample and so must operate at much higher speeds than a conventional small computer; the alternative is a frequency analysis and regeneration. This is why filtering as such is rarely carried out in real time on a normal computer. The major advantage of digital filters is that they can be controlled easily by computer since only sample timing needs to be altered, unlike analogue filters which require switching of resistors or capacitors.

High order filters (eighth order) tend to be digital, whereas lower order filters are more likely to be analogue. This reflects the difficulty in setting up and 'gagging' many potentiometers or resistors to give high order analogue filters. Conversely, due to operating speed limitations, it is not common to find digital filters that operate at very high speeds.

## 13.5 Bandpass or resonant filter

If a vibration of interest occurs at, say, 500 Hz we can either use a 500 Hz resonant filter, i.e. a single degree of freedom system with low damping, or we can use a bandpass filter set to either side at perhaps 400 Hz high pass and 600 Hz low pass. Since both will give transmission of 500 Hz and will cut out low and high frequency noise, why should we use one rather than the other?

The resonant filter gives a roughly symmetrical cutoff and its performance is usually stated in terms of a $Q$ or dynamic magnification factor of the circuit. A $Q$ of 50 is achievable, whence the effective bandwidth of such a filter to the half-power points would be 2%. A 20 dB reduction occurs by about ±22% change of frequency from resonance and so cutoff performance is comparable with an eighth order bandpass filter.

Figure 13.5 gives a comparison between a bandpass filter and a resonant filter but a simple amplitude check does not indicate the important difference, which is the response to transient vibration. A rough rule of thumb is that a resonant circuit will require of the order of $Q$ cycles of vibration at resonance frequency to build up to full amplitude, and thus if only three cycles of vibration occur there

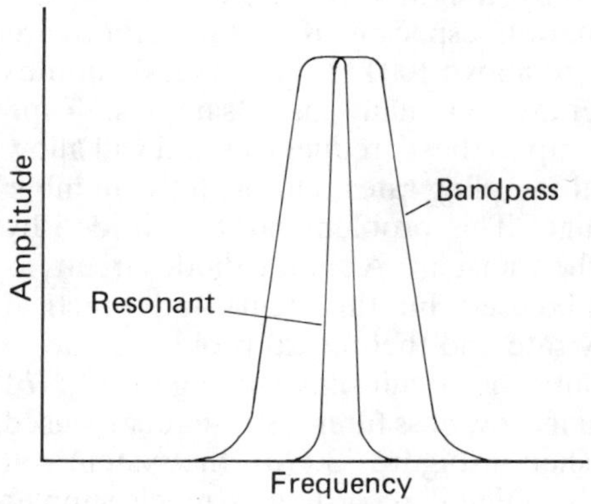


**Figure 13.5** Resonant and bandpass filter responses

will be little build up of vibration towards the 'correct' amplitude. Conversely when a vibration stops suddenly there will be an appreciable number of cycles (of the order of $Q$) before the filter output dies away. In contrast a bandpass filter will respond almost immediately and will decay immediately unless a very low damped circuit such as a Cauer is used. A short transient effectively increases the required bandwidth.

Once again it is important to consider the engineering requirement when making the choice between the two types of filter, since a 'steady' component is sometimes required or, alternatively, the exact position and length of the transient is important. Also, if there is a likelihood of slight variation in the vibration frequency the resonant type is unsuitable since the signal will fall outside the very narrow peak.

## 13.6 Bursts of vibration

In some vibration work, such as a monitoring for failure, the signal of interest is a fairly small signal of high frequency vibration which only occurs in irregular, shorts bursts or transients. A single transient may last less than 1 ms with a frequency of about 30 kHz, and transients may only occur once per second on average.

Direct frequency analysis of the signal is of little use since power levels from the transients are low because they occur for such a short time and there is often a steady background noise at the 'carrier' frequency. Visual inspection of the signal is very effective unless the signal is submerged by other frequencies, usually lower; a bandpass filter set 10% either side of the 30 kHz carrier is suitable.

After filtering to leave only the 30 kHz, there remains the problem of volume of information, especially if digital methods are used; 30 kHz infers sampling at above 100 kHz and several minutes of monitoring would generate too many data samples. Tape recording will easily function up to these frequencies and will allow reduction of speed to normal sampling rates, although the number of samples would still be high. This problem can be avoided by generating the envelope of the transient. A simple diode circuit, as shown in Figure 13.6(*a*) can be used, but this incurs both practical problems of choice of decay rate and theoretical problems due to the non-linearity of the output. The circuit shown in Figure 13.6(*b*) gives simple rectification and if a low pass filter is subsequently used to give the signal shown dashed in Figure 13.6(*c*), the system will respond rapidly and linearly. Digital recording is much simpler because the sample rate corresponds to the pulse length, rather than

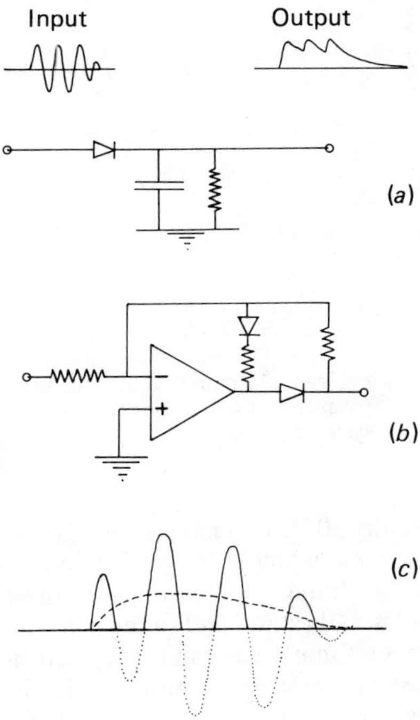


**Figure 13.6** (*a*) Simple diode circuit. (*b*) Improved rectifier circuit. (*c*) Resulting rectified (full line) signal and smoothed (dashed line) signal

to the oscillatory frequency, and can be reduced to about 5 kHz instead of more than 100 kHz.

## 13.7 The Hilbert transform

There is a good deal of interest in applying the 'Hilbert transform' to vibration signals, especially where digital processing is being used.

The definition in the form of an integral over all values of $u$ of $[x(u)/\pi(t-u)]\,\mathrm{d}u$ is not at all helpful for understanding of the process.

In the frequency domain, the effect of the transform is somewhat clearer as it turns out to have the effect of a 90° phase shift. In its effect it is similar to differentiation (or integration) as far as phase is concerned but, unlike either, does not alter amplitude and thus is simply equivalent to multiplication by $(-j)$ for +ve frequencies (and $+j$ for −ve frequencies).

Most electrical AC work uses ideas of rotating vectors (phasors), as shown in Figure 13.7 where a vector representing a voltage or

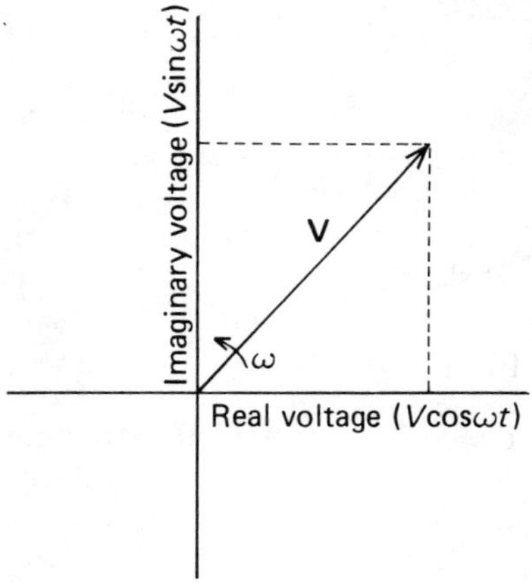


**Figure 13.7** Rotating voltage vector (phasor) as used for electrical AC systems

current rotates at frequency $\omega$ (usually 50 Hz). What we measure is the instantaneous 'real' voltage $V \cos \omega t$, which is the projection on the real ($x$) axis and there is also an 'imaginary' voltage $V \sin \omega t$ which is not used. AC networks are 'steady state' and so the rotating vector **V** is of constant length, but the same ideas can be applied to vibration with the complication that the rotating vector is varying in amplitude with time. We still observe the 'real' projection on the real axis, the $\cos \omega t$ part, but can synthesise the 'imaginary' part on the other axis with the Hilbert transform.

Figure 13.8 shows the effect for a burst of vibration with the vibration vector tracing out a bulging helical or screw path as time progresses. What is observed is the 'real' part projected onto one axis. The interest in the signal is not usually in the frequency, since

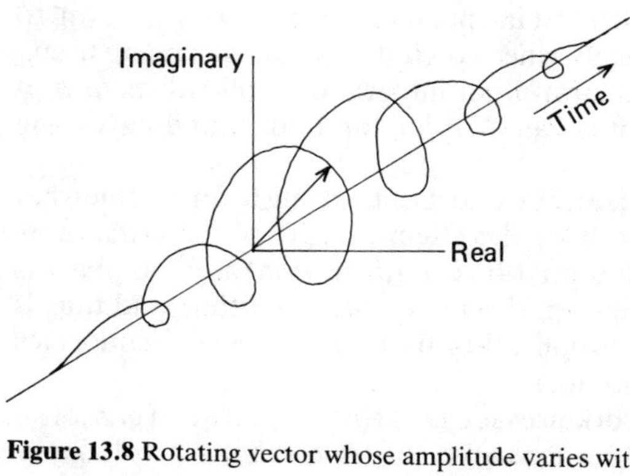


**Figure 13.8** Rotating vector whose amplitude varies with time

this is normally known as a resonance of the system or has been selected by filtering, but, rather, is in the amplitude of vibration.

The analogue approach to a signal of this type is to generate the envelope, and the Hilbert transform is essentially the digital equivalent since it generates the 'real' and 'imaginary' parts of the signal so that amplitude can be extracted. In the process there are certain mathematical conditions that should be met since there are inherent assumptions that the vibration is 'well-behaved', implying no sudden changes of amplitude, phase or frequency. Fortunately this is so in normal vibrations.

From the users' point of view the analogue method of extracting the envelope is very much simpler and cheaper and greatly reduces the number of points to be sampled. The corresponding advantage of working with the information in the digital form, and hence using the Hilbert transform, is that it can be carried out at leisure on the digitised information. The end result is a measure of the modulation of the high frequency signal and, in the case of amplitude modulation, there is no significant difference between the methods.

Frequency modulation of a signal is less common. It is not detected by simple analogue envelope techniques but with the Hilbert approach it appears as a phase variation. If, however, phase or frequency variation of a wave is suspected, it is probably clearest to subtract the basic carrier wave from the signal (described as line elimination – see Section 13.8) and view the residual signal which will clearly show up amplitude or phase changes.

## 13.8 Line elimination and signal regeneration

One method of 'improving' vibration signals is the digital equivalent of a very good 'notch' filter, and this has the advantage that it can be applied to recorded signals after testing so that pre-setting of filter frequency is not required. The technique is exceptionally useful where a large, steady vibration is present in a line spectrum but where the signal of interest is the variation from the steady signal. An example of this can occur in a gearbox where there will normally be a large steady component at once per tooth frequency (and harmonics) but the interest lies in variations such as intermittent change in amplitude, phase or wave shape.

The technique is straightforward but longwinded. The vibration signal is frequency analysed, after averaging if relevant, to give phase and amplitude; a typical amplitude plot derived from Figure 13.9(*a*) could appear as shown in (*b*). The known, expected, lines are deleted from the plot and the spectrum left over is used to

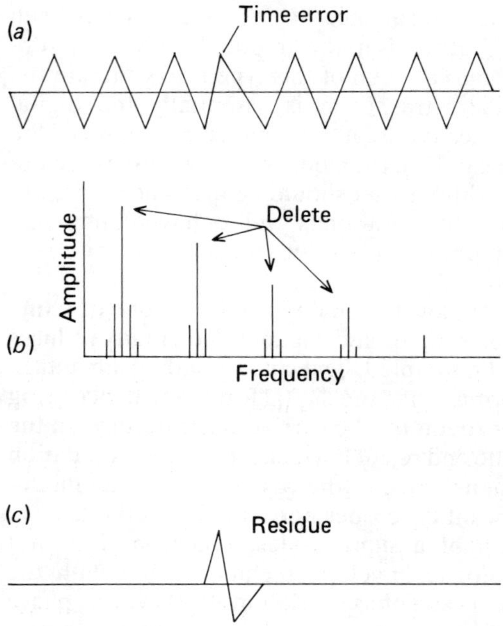


**Figure 13.9** (*a*) Vibration signal with timing error on one cycle. (*b*) Corresponding Fourier analysis. (*c*) Residual signal after standard main lines have been deleted from Fourier analysis and signal regenerated

regenerate a time signal. As the regeneration process is the exact inverse of Fourier frequency analysis it is not difficult to achieve, although the phase information must be used. The end result is theoretically the same as that obtained if perfect notch filters with no phase distortion had been inserted for tooth frequency and its harmonics. Figure 13.9(*c*) illustrates the result that would be obtained if the original signal had a slight timing error on one tooth; comparison with the original signal clearly shows how the variation from normal is emphasised.

In this example a similar result could have been obtained by time averaging at once per tooth frequency and continuously subtracting the average from the vibration signal. In general, working in the frequency domain is rather more flexible and adaptable if there are several sources of vibration in the signal. Irregular firing on an IC engine is a standard case where displaying variations from the average is more instructive than analysing the original signal.

Chapter 14

# Condition monitoring

## 14.1 The problem

The objective in all condition monitoring of machinery is to obtain the earliest possible warning of damage or malfunction and to assist diagnosis of the cause of the malfunction. In much of conventional industry the advantage of having an 'early warning' system is to allow shutdowns to be scheduled at convenient times. Sometimes the requirement is more critical, as with helicopter gearboxes or aircraft engines where both secondary damage and danger to life is to be avoided.

Two areas dominate monitoring, bearings being the major problem and gears rather less important. As yet there are no absolute answers to the problem and so a variety of techniques are used.

Initially an accelerometer was fitted to a convenient bearing housing, a simple measurement of total vibration power was made and an arbitrary level of power was set on the basis of experience. This approach is still possible for simple systems where a single shaft runs in a plain bearing and the most likely trouble is an out-of-balance. For the more typically complicated problems a whole array of methods can be and is used with no one method being wholly reliable.

## 14.2 Frequency analysis

The first major improvement was to carry out a frequency analysis on the vibration and this gave much earlier indication of problems in a complicated installation. Figure 14.1(*a*) shows a possible frequency analysis under 'good' conditions, while in (*b*), there is a major increase in the vibration at one frequency, although the total power has not altered significantly.

Trouble may be indicated by an increase in power at one characteristic frequency or it may be shown by an increase in the modu-

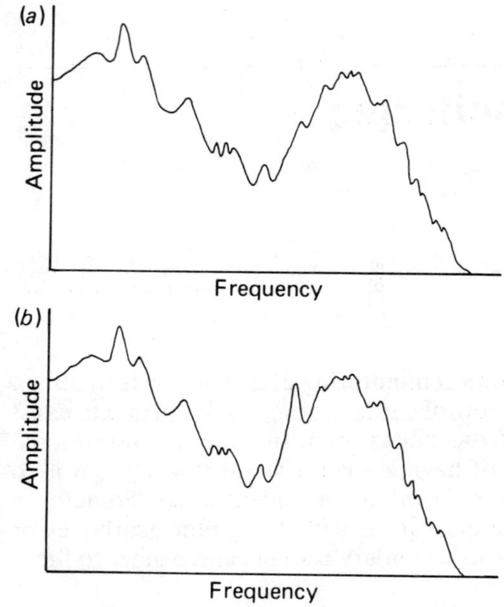


**Figure 14.1** (*a*) Spectral analysis under normal conditions. (*b*) Analysis with trouble at one frequency

lation and consequent increase of the sidebands. Frequency analysis is very good at detecting increases in steady vibration such as out-of-balance effects but is less sensitive for intermittent faults such as, say, one damaged tooth in 100. Figure 14.2 shows diagramatically an increase in once per revolution vibration (from *a* to *b*) which would be detected easily by frequency methods, while the corresponding single tooth damage in Figure 14.3(*a*) would not give

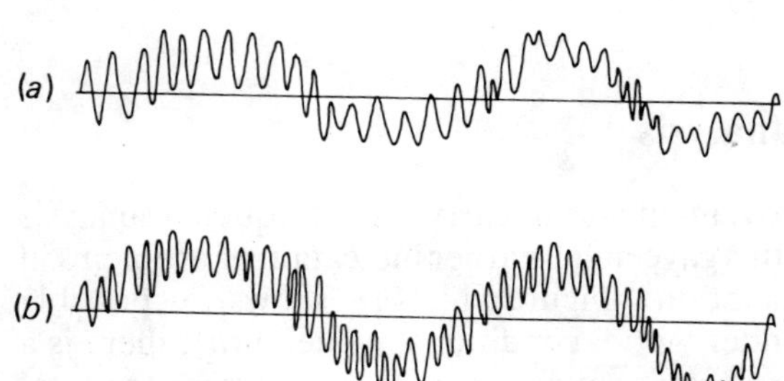


**Figure 14.2** Vibration traces showing increase in low frequency component but negligible increase in high frequency amplitudes

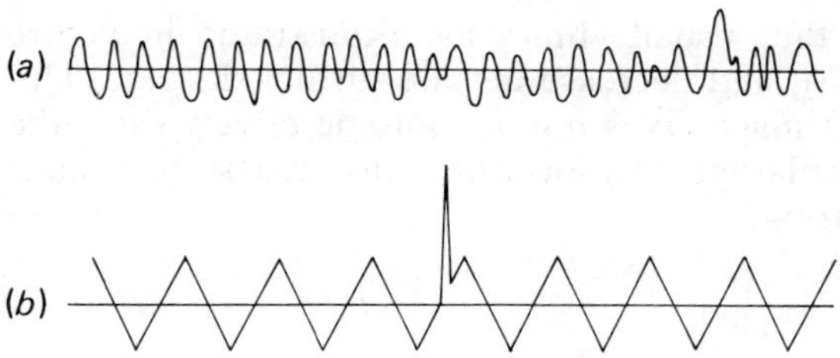


**Figure 14.3** (*a*) Typical vibration from damage on a single tooth. This would give little extra power at any single frequency. (*b*) Vibration signal with isolated narrow peak, which could be noise or trouble

much increase in either total power or any single line. Existing background noise will usually swamp any small increases due to the single tooth. As the second fault is obvious at a glance on the time signal, while the first is not, different methods must be used for intermittent faults.

A major problem with the use of frequency analysis (sometimes called machine signature) is that there is no absolute level that indicates a fault and so it is necessary to build up a back log of experience in order to know what are normal levels and what are trouble levels. Machines vary appreciably and it can happen that the 'trouble' level of vibration on one machine is less than the 'normal' level of vibration on an apparently identical machine. It is necessary to monitor a particular machine throughout its life and watch for 'trends' in the signature. The assumption is that the amplitude of a particular frequency line is 'monotonic', meaning that it increases

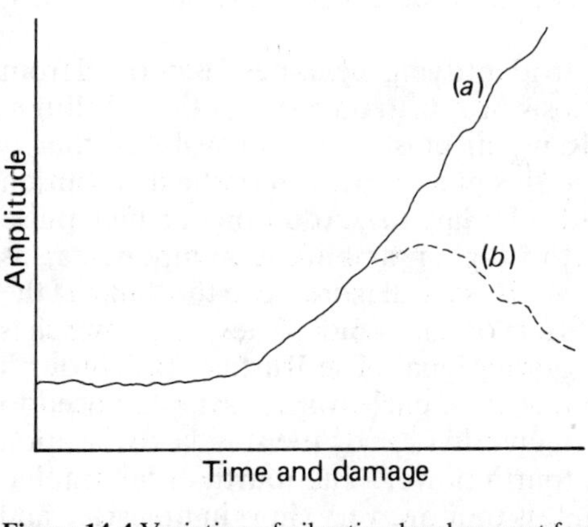


**Figure 14.4** Variation of vibration level or crest factor with time. (*a*) increases continuously but (*b*) is non-monotonic and decreases as the damage spreads

with damage (and thus usually time) as is illustrated in Figure 14.4(*a*), not as in (*b*), and decrease despite further damage. Unfortunately, some damage gives non-monotonic effects since the amplitude of a particular line in a spectrum can decrease as damage spreads round a bearing.

## 14.3 Crest factor and kurtosis

The type of signal shown in Figure 14.3(*a*) is easily assessed by eye but is not easily detected by frequency analysis. The simplest criterion used is the 'crest factor', which is the ratio between the peak value of the signal and the rms value. A steady sine wave has a value of $2^{1/2}$ and so a crest factor of above, say 3, suggests an irregularity in the signal. This approach is very efficient where there is no previous history of vibration and hence no possibility of detecting trends in amplitude of spectral lines or peaks.

There are two major problems with the 'crest factor' approach. The first occurs because a peak value is compared with a background rms; in the case of a gear mesh the background is due to the regular tooth meshing and thus if an exceptionally good gearbox is accidentally manufactured there will be a very low level of rms signal. Even a minor defect will then give a high crest factor. The second problem is indicated in Figure 14.3(*b*) where a visual inspection might suggest that the high peak is an isolated noise pulse and so may be ignored, whereas a standard simple computer routine would indicate a high crest factor. Time averaging of the signal would eliminate such peaks if they were irregular, but this is not always possible.

An alternative approach to identifying 'peakiness' is derived from statistics and is called Kurtosis or fourth moment of the amplitude probability distribution. Random noise has a probability that is Gaussian and thus with a mean signal value of zero the distribution will be as sketched in Figure 14.5 line (*a*). Addition of a high pulse increases the amount of the higher amplitude components, as shown in Figure 14.5, line (*b*). This small increase in the 'tails' of the distribution has very little effect on the standard deviation, which is the same as the rms value of the signal. The Kurtosis is the fourth moment of the distribution, that is, each voltage sample raised to the fourth power, averaged, and divided by the standard deviation which is also raised to the fourth power. The fourth order emphasises peaks in the system unless they are very short in timescale and so is better than a simple crest factor for ignoring noise, provided

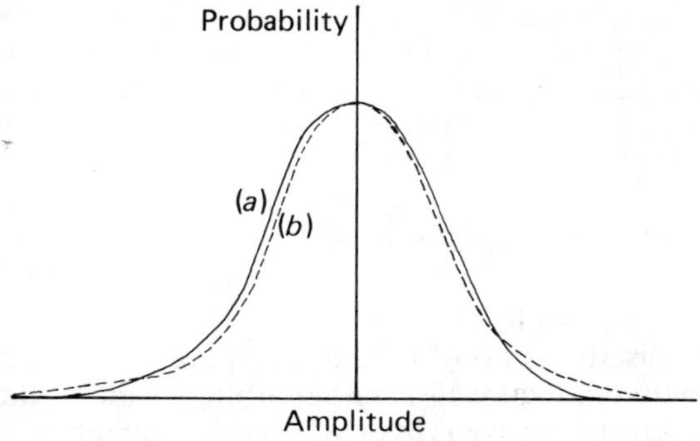


**Figure 14.5** Statistical distribution of vibration signal (*a*) with random noise and (*b*) with extra high amplitude peaks as in 14.3(*a*)

that the noise is of short duration. The problem of a low rms with a good gearbox still remains.

A more sophisticated version of the Kurtosis splits the raw vibration signal into several broad frequency bands and checks each band separately for its Kurtosis value. Random noise gives a value of 3, and a value of around 5 suggests damage but, unfortunately, both crest factor and Kurtosis are not necessarily monotonic and for a rolling bearing may decrease as damage progresses and spreads round the bearing. Also the Kurtosis of a regular sine wave has a much higher value than that of random vibration and thus may cause problems.

## 14.4 Impulse detection

In theory a rolling bearing should produce high frequency impulses if there is a defect such as a small pit on the track. A medium size bearing at electric motor speeds has a rolling velocity of the order of 1 m/s so if it encounters a defect of width 100 $\mu$m we might expect a pulse of length 100 $\mu$s and rise and fall times of the order of 20 $\mu$s. Any method that can detect pulses directly would offer a possibility of early detection of damage.

The 'shock pulse meter' method uses an accelerometer with low damping so that it 'rings' if a pulse arrives. The natural frequency chosen is around the 30 kHz region where there is normally relatively little vibration from other sources in the machinery. The 'ring' from the accelerometer can then be further filtered and amplified using a tuned electrical circuit. The equipment is simple and robust

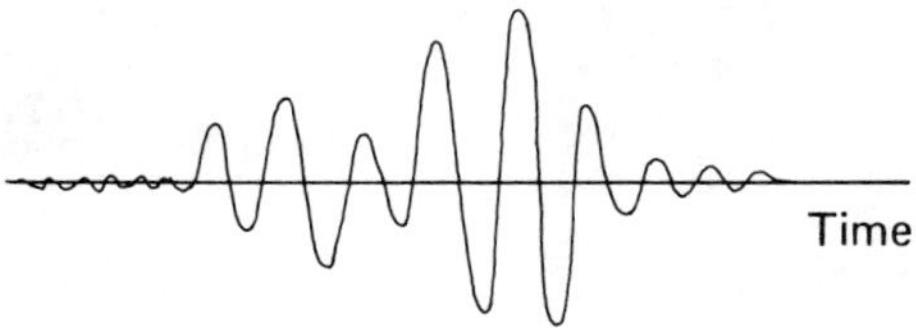


**Figure 14.6** Typical 'impulse' observed at accelerometer

and gives reasonable discrimination for bearing failures, although inevitably the setting of 'failure' levels is rather arbitrary and trend analysis is desirable. An alternative is to use an accelerometer with a very high frequency range and to carry out all the resonance filtering electronically. This variant does not restrict the detection to a particular accelerometer mounting frequency.

Experimental observation of the impulses obtained from a bearing show that they bear little similarity to the expected shapes. An 'impulse' is likely to be of the shape shown in Figure 14.6 with a build up of vibration over several cycles and subsequent decay. The frequency involved is not associated with the very high frequency accelerometer but, rather, is that of the individual balls or rollers vibrating on their contact stiffnesses. This ball resonance effect is the only method by which vibration can travel easily from the inner to the outer ring of a bearing and so is a very convenient mechanical filter for monitoring. The frequency varies somewhat and the resonance of each installation should be identified by frequency analysis so that the bandpass filters can be set correspondingly.

Once the impulses have been obtained and all other vibration filtered out, enveloping or rectification, as discussed in Section 13.6, can be used to reduce the amount of information. Subsequent analysis is almost a matter for individual choice since whether peak-height, total power or peak repetition rate is important is open to argument and preferably should be tested experimentally for a given application.

## 14.5 Averaging

Time averaging techniques for improving the accuracy of the signal and reducing quantity of information are useful for any fault that is reliably regular, but not for transient or irregular ones. The first category would include any gear tooth defects, consistent misfiring in an engine, pump plunger sticking, etc.

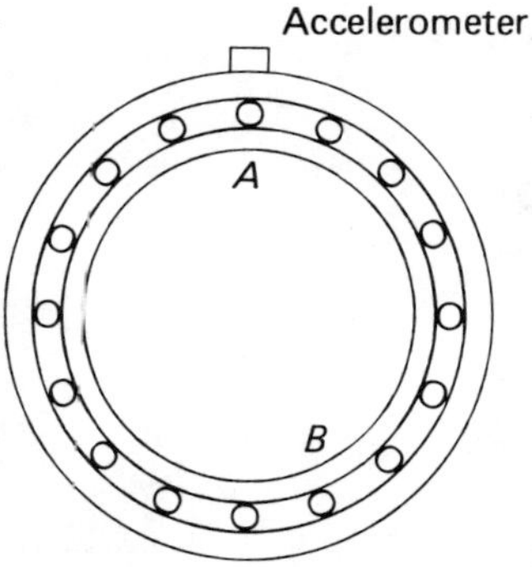


**Figure 14.7** Sketch of bearing. Observed pulses when damage is at *A* will be larger than from *B*

Unfortunately, rolling bearings do not repeat exactly regularly since the cage is not revolving at an exact integral speed ratio of the inner rotation speed. The interval between pulses is consistent if an angular contact bearing is used, as each ball will pass over an inner race defect. The amplitude of the observed pulse will depend on whether the impact occurs near the accelerometer or well away, as shown at *A* and *B* in Figure 14.7. A conventional bearing will only have contact on one side, over an arc of about 130°, and coincidence of ball and defect outside this area will thus not give an impulse. The pattern of vibration observed will be as shown in Figure 14.8 with some 3 or 4 pulses per revolution disposed over the arc of contact. In this case, averaging after enveloping will give the shape shown in Figure 14.9(*a*), the shape of the arc of contact modified by the transfer attenuation between balls and detector. Correlation would give the slightly clearer signal shown in Figure 14.9(*b*) with only exact intervals between pulses being possible.

Complications arise when there is no longer a single defect since as the trouble, and hence the pulse generation, spreads round the bearing the average degenerates to a steady signal. The correlation will also degenerate since pulses will be spaced corresponding not only to ball spacing but also to distances between defects.

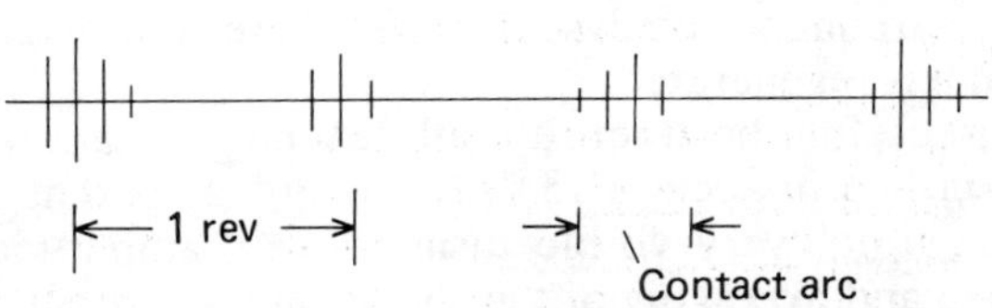


**Figure 14.8** Observed pattern of pulses with batches at 1/rev spaced 1/ball contact apart

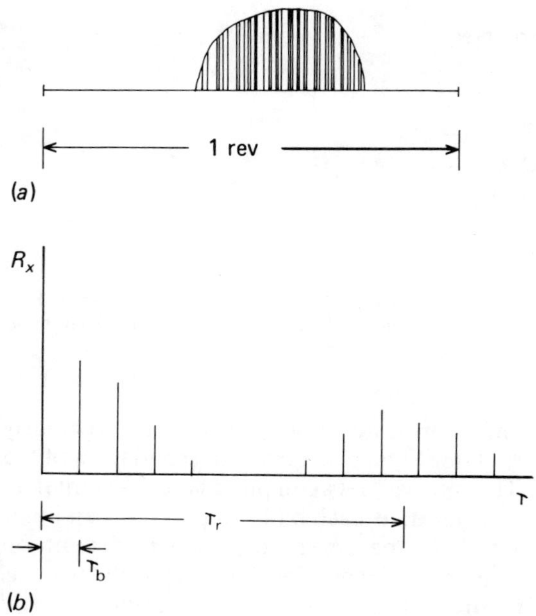


**Figure 14.9** (*a*) Pulses during a revolution will only occur within the contact arc position with amplitude roughly proportional to load. (*b*) Correlation of vibration signal. In general there will not be a correlation at exactly 1/rev since that will not correspond to an integral number of ball intervals. $\tau_r$ is the time for 1 revolution and $\tau_b$ the ball pass period; $R_x$ *is the resulting correlation*

## 14.6 Line elimination

Gearboxes usually have a powerful regular signal at once per tooth and as this signal is large it is worth while eliminating this part of the signal to make it easier to detect small peaks from trouble sources.

Although in theory this can be done by averaging at 1/tooth and subtracting the average signal from the original, the method is rarely used since extra synchronising signals at 1/tooth are needed; this approach is, however, inherently faster than the alternative. Normally the signal is frequency analysed, the lines are subtracted and the residual signal is regenerated.

The method is most useful when there are subtle timing variations in the signal, as described in Section 13.7. Its advantage is that a phase variation that is not very visible turns into an amplitude variation. Enveloping and averaging of the signal can then greatly reduce the amount of information stored while retaining the character and position of the defect. After the lines have been removed, crest factor or Kurtosis assessment is more likely to be effective. The removal of lines from the frequency spectrum can be thought of as equivalent to the use of a perfect notch filter.

Chapter 15

# Shock testing

## 15.1 Introduction

For the majority of civil industrial requirements the endurance testing specifications do not present major difficulties. A specification of $\pm 3\,g$ sinusoidal shocks for perhaps 10 000 cycles at 50 Hz is easily achieved and the typical 24 hours at $\pm 1\,g$ is also easy. A medium size electromagnetic vibrator giving about 1000 N force (220 lbf) allows a total mass of some 30 kg at 3 $g$ acceleration. Part of this may have to be a guidance system to hold the item and ensure that it moves in a straight line, but even after allowing some 10 kg for this, the payload can be about 20 kg, more than enough to cover most contactors, electronics units, etc. It is sometimes necessary to balance gravity effects and support the weight with soft elastic cords (bungee) when testing vertically, or to make a slip table for horizontal tests.

Measurement of acceleration is usually easy since reliable calibration is available at $\pm 1\,g$, and since the vibration is steady sinusoidal it is possible to check by measuring the amplitude at low frequencies. At 10 Hz, $\pm 1\,g$ corresponds to an amplitude of $\pm 2.5$ mm, or 0.1 inches. It is safe to assume that the calibration remains linear between 1 $g$ and 3 $g$.

There is no exact definition of what constitutes a shock but we would normally consider a single impulse loading to be a shock, whereas a steady sinusoid is called a vibration. In practice neither may be relevant since the excitation is sometimes a short burst of vibration, mostly at a single frequency. When, as in some earthquakes or machine drive starting transients, there are 5–10 oscillations we cannot rigidly classify the problem.

The problems in shock or vibration testing arise when payloads or accelerations become high; $+/-3\,g$ on a mass of 500 kg can be achieved, but an electromagnetic exciter giving 15 000 N is large and expensive. Regardless of size or power it is difficult to get electromagnetic vibrators to exceed 100 $g$, even with no mass on the exciter, because the weight of copper windings and former limits the acceleration.

## 15.2 Large loads

Large forces can be achieved relatively cheaply by using hydraulic vibrators instead of electromagnetic exciters. A rough rule of thumb is that a hydraulic system should be considered once the required force exceeds half a ton, 5000 N. The limitations lie in the dynamics of the system when frequencies rise.

An electrohydraulic servo valve typically works at 200 atmospheres, 20 MPa, and can give a flow of 20 l/min (20 $in^3$/sec) with a natural frequency, heavily damped, of up to 200 Hz. A force of ±1 tonne, 10 000 N, requires a ram area of 500 $mm^2$ and thus a ram of OD about 30 mm. The flow limits at 0.3 l/s and so the maximum ram velocity is limited to 0.6 m/s. The limit at 10 Hz, if we assume sinusoidal motion, is $0.6 \times 20\pi$, giving only $\pm 4\,g$ acceleration and ±10 mm aplitude. By 100 Hz the acceleration has risen to a theoretical $\pm 40\,g$, although compressibility effects in the oil will be starting to reduce performance, and after 150 Hz the valve flow reduces. Increasing the force level to ±20 000 N doubles the ram size and so halves the velocity and acceleration. Fitting two valves working in parallel is difficult in a limited space but doubles the available acceleration.

## 15.3 High acceleration

As the limiting acceleration achieveable with electromagnetic (moving coil) exciters is at most 100 *g* and hydraulics cannot approach this figure, it is necessary to find alternative methods for accelerations of 100 *g* and above.

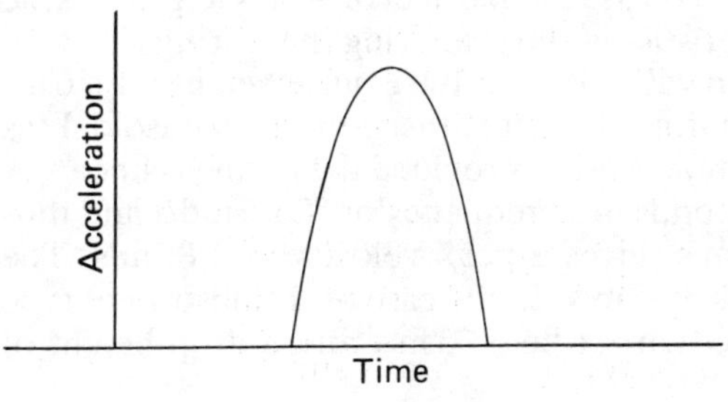


**Figure 15.1** Required impulse pulse shape

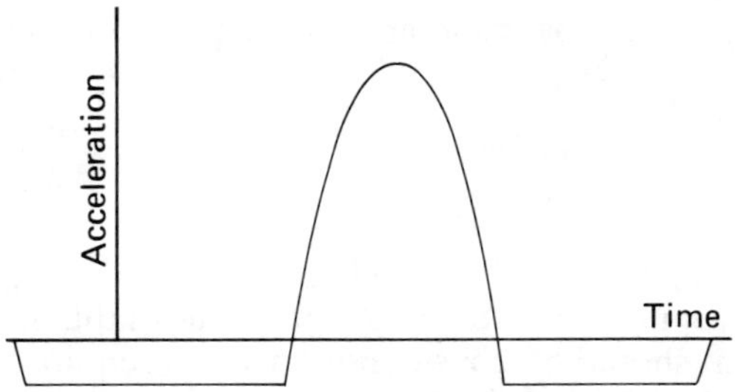


**Figure 15.2** Actual pulse shape to give total velocity change of zero

Specifications usually call for a single impulse with a half-sine wave shape, as shown in Figure 15.1. This acceleration pulse is only possible if there is a finite change of velocity and as, sooner or later, it is advisable to have the system stationary, the actual pulse is likely to be as shown in Figure 15.2 with as much area below as above.

The methods used for high accelerations are essentially mechanical. Cams are not common but can be used and are surprisingly effective; it is of interest that the valves in a car engine, cam driven, normally run up to 500 *g*, or in racing engines, over 1000 *g*.

Dropping the system onto a controlled 'impact' is the most commonly used and effective method. The 'crash' can give either a purely absorbent stopping of the carriage or a semi-elastic bounce where there is reversal of the carriage motion. Figure 15.3 is a sketch of the system; velocity control is achieved by adjusting the dropping height. In the case of the absorbent impact a spike indents into some form of honeycomb of plastic which will distort consistently, and the shape of the spike is adjusted to give the correct pulse shape. This is not easily achieved. The alternative is to use a spring which must have a very high response speed so that spring surge effects do not distort the resisting force and so result in a poor pulse shape. The simplest and fastest reacting springs are conical disc springs and these can be used in parallel or series to give a wide range of spring characteristics without altering the test rig.

A typical specification will call for a 100 *g* pulse height for a 6 ms duration. The motion during contact is theoretically sinusoidal (as required) if a spring system is used, provided the spring is linear. A half-sine of 6 ms corresponds to a frequency of 523.6 rad/s and thus an acceleration of 981 $m/s^2$ gives a peak velocity of 1.87 m/s. The change for a half-cycle is 3.75 m/s. In the case of the absorbent type of test the impact velocity must be 3.75 m/s and a drop height of 0.72 m is required.

An elastic bounce is never fully elastic and trial is needed to

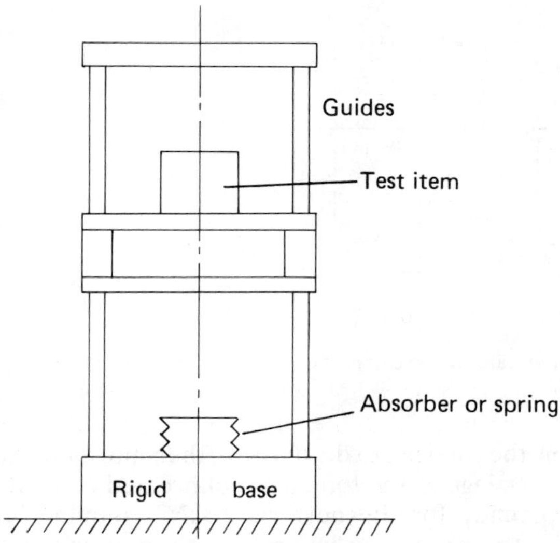


**Figure 15.3** Sketch of impact system

ascertain the correct height. If the rig is well designed a coefficient of restitution of 0.6 might be obtained and the contact velocity would then be 2.34 m/s, giving a drop height of 0.28 m. Accelerations of higher $g$ levels are often shorter in duration and so drop heights do not necessarily increase since drop height is proportional to the square of the product of peak acceleration and pulse time. Spring compression is of the order of 2.34/523.6, which is 4.5 mm.

The most likely source of problems in rig testing is deflection within a non-rigid item under test. This happens when equipment is not held rigidly in its case, the case decelerates first and the contents some time later with a delay in the reaction force which distorts the motion of the carriage. The simplest solution is to increase the mass of the carriage (and the corresponding deceleration forces) to dominate the equipment forces.

Tests requiring a single impulse can equally well be done by the absorbent or bounce approach but if the specification calls for a large number of shocks, say 1000, the use of a fresh absorbing honeycomb or foam would be very time-consuming. The advantage of the spring bounce system is that it can easily be automated.

A standard moving coil oscillator can be used with a waveform, as shown in Figure 15.4, to raise the carriage and hold it against an upper stop. The length of the downward pulse is governed by the time taken to drop, impact and bounce. The coil moves downwards

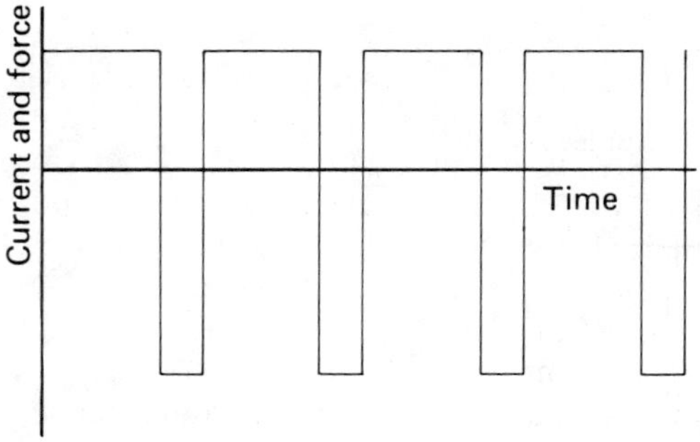


**Figure 15.4** Waveform for vibrator to raise impact carriage and then catch it after impact

out of the way to allow the carriage to drop freely, then moves back upwards to catch the carriage at the top of its bounce and to lift it back against its stop ready for the next cycle. DC coupling is required but the vibrator need only be sufficiently powerful to support the weight of the carriage. Coil travel can be a problem if the drop height exceeds the vibrator stroke, although a lever or additional elastic cord downforce can be utilised to reduce stroke length.

An air cylinder can be used as the support member and will move downwards fast enough, but large valves and ports are needed to ensure that it quickly returns to catch the bounce.

## 15.4 Measurement

Accelerometers are reasonably linear devices over small ranges but over large ranges will give up to 10% non-linearity, so it is not very satisfactory to calibrate an accelerometer at $\pm 1\,g$ and expect accurate results at $1000\,g$, however accurate the scaling range resistors on the electronics.

One possibility is to use a vibrator to its limit at $\pm 100\,g$ and measure amplitude with a microscope, but this involves a 10 : 1 extrapolation and hence high errors even if a perfectly sinusoidal motion occurs.

A better solution is to use a circuit with a high frequency response op-amp to integrate the acceleration signal to velocity.

The integration process is only dependent on passive components and the time constant of the circuit can easily be determined, accurate to 0.1%, by testing with a sinusoidal input so that there is no loss of precision in the integration process. Calibration of velocity change can be carried out by making a separate accurate measurement of impacting speed (and bounce speed if there is a bounce) using accurately spaced photocells to start and stop a timer. Gravity acceleration of the carriage means that the speed will not be steady; correction for this effect can be made by estimating contact velocity from average velocity between the timing points. A slightly more sophisticated measurement can be made by having a single fixed photo-diode and detector and attaching to the carriage a strip with holes drilled at about 2.5 mm intervals. Timing the string of pulses gives both the velocity and acceleration before and after the impact.

Comparison of the velocity change with that from the integrated accelerometer signal will give an accurate velocity, and hence acceleration calibration. The limits on the system are set by transient digitisation accuracy, not important if 12-bit resolution is used, and uncertainty due to lack of linearity. The latter is not large as the velocity change is controlled mainly by the high acceleration part of the impact.

Chapter 16

# Automation of testing

## 16.1 Introduction

A wide range of approaches to testing is possible, irrespective of whether the object of interest is a machine tool, car suspension or oil rig. One extreme is the basic 'hand' method with a small vibrator using a slow sine sweep, a hand-applied accelerometer and hand plotting of results. Capital outlay is of the order of hundreds rather than tens of thousands of pounds but a set of test results will take 2 days rather than 2 minutes.

The other extreme is most impressive, with several large exciters and scores of permanently attached accelerometers multiplexing to a high speed computing installation. Such a set up, complete with software, is expensive but will produce beautifully displayed results within minutes. Natural frequencies, mode shape plots or animated sketches can be provided in contrasting colours for viewing or can be plotted out.

It should, however, be noted that both approaches essentially derive the same basic information and should come to the same conclusions.

## 16.2 Relative speed

At first sight the 'fully automated' system (taking 2 minutes instead of perhaps 10 hours by hand) is at least a factor of 100 faster, and sales literature may suggest that 100 tests can be done in the time formerly taken by one.

The reality is much less impressive due to the set up times involved for each machine. Accelerometers must be fixed firmly in position by screwing or gluing; an exciting vibrator must be mounted and connected or a hammer positioned; and all the information on exciter and detector positions must be typed into the computer. This setting up time is comparable to the time taken to determine one or two mode shapes by hand.

The major speed advantage of the automated set up does not occur due to shorter testing times, but does occur when the processing of results is complicated and hence time-consuming. If four resonances are interacting between 400 and 500 Hz in a complex system with many degrees of freedom, it would require a great deal of time to plot out a polar response, deduce the frequencies and damping constants and separate out the mode shapes. In this case the computation of the results is very well worth while economically.

In contrast, a typical machine tool self-excited regenerative chatter may occur at a clearly defined frequency with no problems of mode interaction, and it is faster to set up a vibrator and to check and plot the resonant mode shape by hand than it is to attach many accelerometers and type information into the computer.

## 16.3 Mode shapes

Ten accelerometer positions recording response along a structure might give results as shown in Figure 16.1. A computer program would put a 'best curve' through these and display a mode shape as shown.

Unfortunately the actual mode shape may be as shown in Figure 16.2 and even if we use 20 accelerometer positions we will not necessarily get a good representation of the 'fine detail' of the mode shape simply because we have automatically instructed the computer to put the best smooth curve through the points and it will, if necessary, ignore points since it regards variation from a smooth curve as 'noise' on the measurements.

Detailed checking of a complicated mode shape, of the kind that might be encountered on a machine tool, may require measurement of over 100 points and this is most simply and quickly

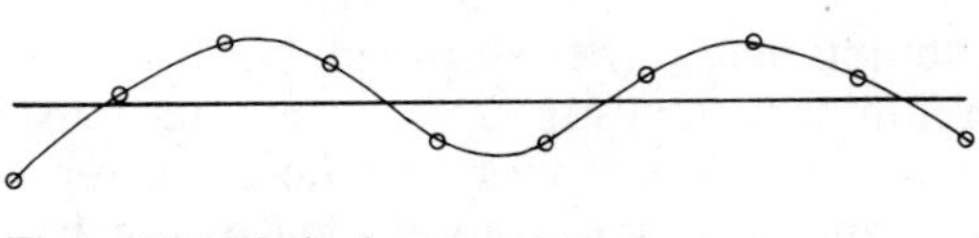

**Figure 16.1** Mode shape as plotted out

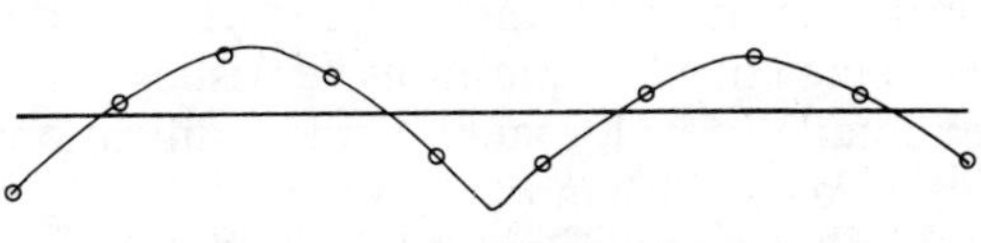

**Figure 16.2** Actual mode shape with 'weak spot'

achieved by having steady excitation and applying a roving accelerometer by hand. Use of a human being, while incurring disadvantages in consistency, has the major advantage that as soon as trouble is suspected, intelligent guesses as to the causes of kinks in the mode shape can speed up testing considerably.

The combination of relative advantages and disadvantages discussed in Sections 16.2 and 16.3 leads to a tentative conclusion as to the conditions under which 'by hand' or 'full automation' is technically better. At one extreme, many interacting modes in a narrow frequency band, each having a relatively straightforward mode shape, will be best tested using the 'fully automated' approach. In contrast, a single isolated mode with a very complicated mode shape is better tackled by hand, using a steady sine wave excitation, and hand plotting so that the roving accelerometer positions are determined by the results just obtained. Hand-held pickups are particularly useful for direction variation to determine the direction of zero acceleration at points on a body to locate instantaneous centres of rotation.

## 16.4 Economics

The investment in a fully automated system is high; to the obvious costs of equipment, computers and software must be added the cost of training personnel to use the equipment. Amortisation rates are high and so it is realistic to budget for overhead costs of the order of tens of thousands of pounds per year.

This level of cost is easily justified if the equipment is used regularly but if only used half a dozen times per year, or less, it is more economic to test manually. A medium sized industrial firm which occasionally encounters vibration problems would be wasting money on a full system and so should consider the possibility of employing a specialist vibration consultancy firm for a system with many modes, or of hand operation for isolated modes.

The 'best' solution is sometimes, rather surprisingly, the use of both approaches. A preliminary investigation by an 'in-house' engineer by hand will determine whether or not there are many interacting modes present and will give a good idea of the best positions to place accelerometers. If modes interact, a specialist team can be called in and their time will be more effectively used if the customer already has an idea of what questions he wishes to ask about the modes. Further detailed development, with modifications to the structure, can be carried out by hand.

Much depends on the level of knowledge and understanding of

the customer since it is of no use to a firm to have a full knowledge of mode frequencies, damping and shapes unless management know how to utilise the information.

## 16.5 Training

Long term strengthening of a company's technical expertise depends on engineers who can measure and interpret vibrations. Accordingly, thought should be given to the best method of building up expertise on the problems of structural rigidity within a firm.

From this point of view, full automation of testing is of little help since humans learn more effectively 'through their fingertips'. The physical process of taking a pickup round a structure and plotting the results seems to provide a deeper understanding of the way the metal is deforming than simply looking at mode shapes on a computer screen. Logic says that there is no difference but in practice the hand operation appears to be better.

Once a vibration engineer is experienced in determining mode shapes (after the first few dozen), it is worth while saving time by automation, leaving only some of the fine detail of mode shapes to be done by hand.

## 16.6 Non-linearities

Non-linearities in a structure can be due to bearings, joints, backlash, friction, hysteresis, etc. In theory their presence can be detected by detailed checking of the computer processed results but in practice it is rare for this to be done and so effects of non-linearities will be lost in 'noise' of the system. The supporting theory for random or impulse excitation of a structure and the corresponding transfer functions are dependent on the assumption of linearity. Checking of coherence should be carried out but if coherence is not 1 there is no simple method of determining the cause.

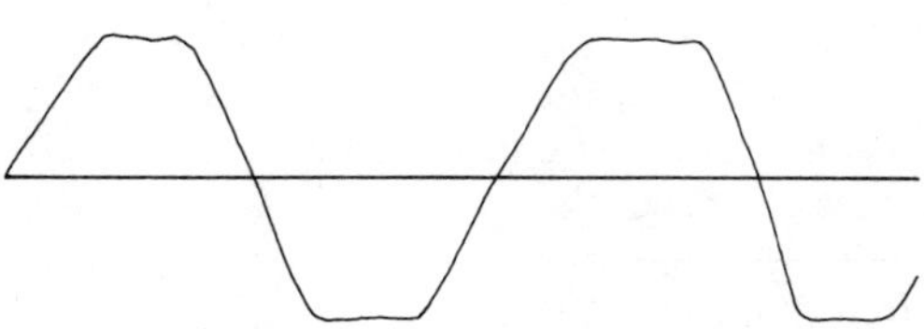

**Figure 16.3** Non-linear saturation effect

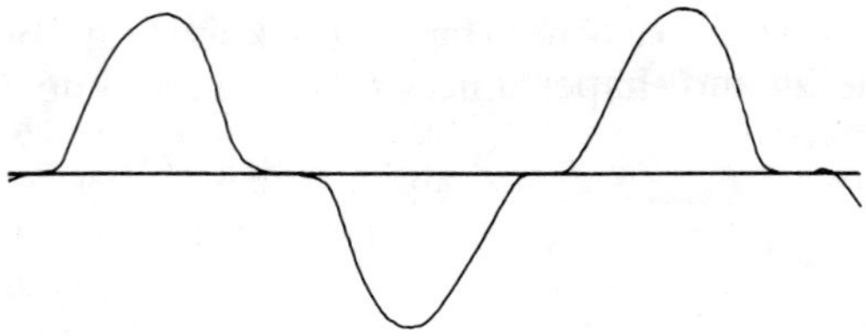

**Figure 16.4** Non-linear play effect

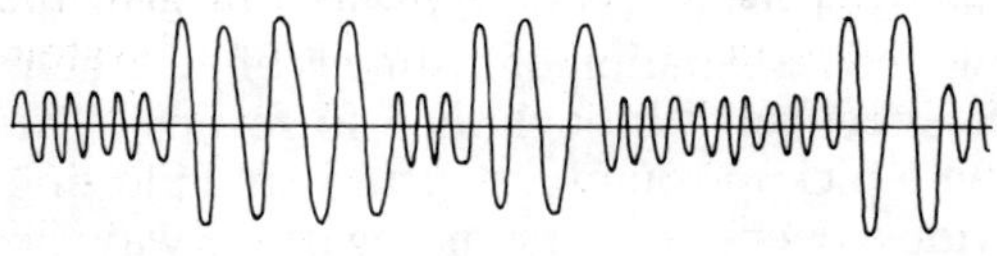


**Figure 16.5** Varying amplitude 'warbling' signal

In contrast, hand checking of vibration waveforms using a swept sine wave input will show up non-linearities immediately on an oscilloscope display. The type of waveform distortion will give an immediate clue to the cause of the non-linearity, as illustrated in Figures 16.3 and 16.4. Occasionally, gross non-linearities will be encountered, giving a signal that fluctuates or 'warbles', as shown in Figure 16.5, due to an irregular bouncing effect. The classic effect of non-linearity in a low damped system is the 'jump' phenomenon, as shown in Figure 16.6; this will show up very clearly with a slow sweep sine wave (preferably increasing and decreasing frequency) but could be missed completely by random or impulsive excitation.

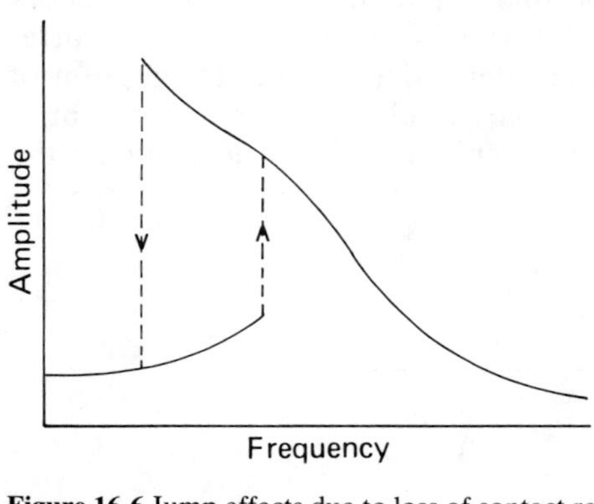


**Figure 16.6** Jump effects due to loss of contact reducing effective spring stiffness

Self-excited vibrations and the similar Mathieu–Hill parametric vibrations due to cyclically varying stiffness give an output without an input and so do not fit into a 'transfer function' approach. Inspection of the raw vibration signal is usually helpful; the case of stick slip is a good example where displaying velocity against time is very informative. Many whirling problems involve orbital motions which are not at shaft rotation speed and so suggest oil whip, asymmetric shaft or hysteretic whirl according to the ratio of shaft speed to whirl frequency.

It is most important to recognise when a system is non-linear since many of the instinctive reactions to linear systems will give the 'wrong' results with highly non-linear systems. Variation of system response with level of steady load is a good indicator of non-linearity. Practical non-linear systems such as rear axles under light loads have been known to give a most deceptive response. Excitation from gear teeth at 500 Hz excited a sub-harmonic resonance at 250 Hz and the resulting vibration generated impacts whose second harmonic, at 500 Hz, travelled easily through the structure. Observation of a 'resonance' at 500 Hz led to the erroneous conclusion that there was a natural frequency at 500 Hz.

The presence of a broad frequency band of vibration, often generated by impacts that may be regular or irregular, is a common indicator of non-linearity and will usually require inspection of the time trace and a model of the system which is rather more complicated than a simple linear model.

The complications of non-linearities suggest that testing should not be fully automated when they are present since the standard routines can be deceptive, and it is thus preferable to utilise human observation as much as possible.

# References

Chapter 1

1. Hartog, Den. (1956) *Mechanical Vibrations*. McGraw-Hill

Chapter 2

1. Cremer, L., Heckl, M. and Unger, E. E. (1973) *Structure Borne Sound*. Springer Verlag, Berlin, p. 280
2. *Mechanical Vibration and Shock Measurements*. Bruel and Kjaer. 92 Uxbridge Road, Harrow, p. 124

Chapter 3

1. Forster, C. P. (1984) Choosing a laser Doppler anemometer. *Chartered Mech. Engineer*. July/August 1984, pp. 30–34. Also DISA, Techno House, Redcliffe Way, Bristol BS1 6NU
2. Licht, T. (1971) *Vibration Measurement by a Laser Interferometer*. Bruel and Kjaer Technical Review, No. 1
3. Schweigelsohn, K. (1963) Entwicklung seismischer Drehfehlermessgeräte mit niedrigen Eigenfrequenzen für die Verzahntechnik, Dissertation, TH Aachen

Chapter 4

1. *Acoustic Noise Measurements*. Bruel and Kjaer, 92 Uxbridge Road, Harrow
2. *Architectural Acoustics*. Bruel and Kjaer, Harrow
3. *Sound Intensity*. Bruel and Kjaer, Harrow

Chapter 5

1. Mudd, G. C., Penning, G. M. and Hilling, N. J. (1983) The application of transmission element measurement to the reduction of airborne and structure-borne noise in gearing transmission, *Proc. Inst. Mech. Engrs., C258/83*
2. Milenkovic, V. and Schmuter, C. (1982) An angular sensor for transmission error inspection, *SAE Int. Congress and Expos. Detroit, 1982, 820054*

Chapter 7

1. Newland, D.E. (1984) *Random Vibrations and Spectral Analysis*. Longman, Essex, p. 120

Chapter 8

1. Broch, J. T. (1975) *On the Measurement of Frequency Response Functions*. Bruel and Kjaer. Technical Review, No. 4

Chapter 9

1. Newland, D. E. (1984) *Random Vibrations and Spectral Analysis.* Longman, Essex, Ch. 12
2. Randall, R. B. (1987) *Frequency Analysis.* Bruel and Kjaer, Harrow

Chapter 10

1. Randall, R. B. (1987) *Frequency Analysis.* Bruel and Kjaer, Harrow

Chapter 11

1. Randall, R. B. (1987) *Frequency Analysis.* Bruel and Kjaer, Harrow

Chapter 12

1. Newland, D. E. (1984) *Random Vibration and Spectral Analysis*, Longman, Essex, p. 210
2. Keendey, C. and Pancu, C. D. P. (1947) *Journal of Aeronautical Sciences.* **14**, p. 603
3. Ewins, D. J. (1986) *Modal Testing: Theory and Practice.* Bruel and Kjaer, Harrow

# Bibliography

In the field of instrumentation a bibliography tends to be a roll-call of the literature from the leading firms in the particular field.

For accelerometers and the associated charge amplifiers the list includes Bruel and Kjaer (92 Uxbridge Road, Harrow HA3 6BZ); D. J. Birchall (Finchley Avenue, Mildenhall, Suffolk); Kistler (Whiteoaks, The Grove, Hartley Wintey RG27 8RN); AVL & Vibrometer (Newby Road, Hazel Grove, Stockport SK7 5EE); Endevco (Back Lane, Melbourn, Royston SG8 6QQ); and Environmental Equipment (Fleming Road, Newbury, Berkshire RG13 2DE). Accelerometers from one source are frequently mixed with amplifiers from another.

Non-contact displacement transducers are almost always bought with their associated electronics since the two are matched; there are numerous competing producers. Unfortunately their literature is usually not at all clear. The best known is probably Bently Nevada (Birchwood, Warrington WA3 7PB).

Strain gauge bridges can be obtained as potted units from Radiospares or Creative Instrumentation Ltd (4 Wayside, Commerce Way, Lancing, Sussex BN15 8TA) for moderate performance, but where high performance is required they can be obtained from Vibrometer, Ffylde (49 Fylde Road, Preston PR1 2XQ) or HBM (Schenk Ltd, Station Approach, Bicester OX6 7BZ).

R. R. Randall's *Frequency Analysis*, published by Bruel and Kjaer, is the clearest and most comprehensive book for those who wish to delve more thoroughly into the problems of analysing signals. A companion book also from Bruel and Kjaer by W. Ewins covers, again very clearly, the problems of *Modal Testing*, i.e. disentangling and identifying modes of vibration which are not separated.

Topics such as coherence and the study of fast Fourier transforms are encountered in the field of random vibrations and *Random Vibration and Spectral Analysis*, by D. E. Newland (Longman, 1984) gives the techniques in this area, as well as the basic ideas of what constitutes a random signal.

# Index